OXFORD PHILOSOPHICAL MONOGRAPHS

Editorial Committee

PROJECTIVE PROBABILITY

ALSO PUBLISHED IN THE SERIES

The Justification of Science and the Rationality of Religious Belief
Michael Banner

Individualism in Social Science
Forms and Limits of a Methodology
Rajeev Bhargava

Causality, Interpretation, and the Mind
William Child

The Kantian Sublime
From Morality to Art
Paul Crowther

Kant's Theory of Imagination
Bridging Gaps in Judgement and Experience
Sarah Gibbons

Determinism, Blameworthiness, and Deprivation
Martha Klein

False Consciousness
Denise Meyerson

Truth and the End of Inquiry
A Peircean Account of Truth
C. J. Misak

The Good and the True
Michael Morris

Projective Probability

JAMES LOGUE

CLARENDON PRESS · OXFORD

1995

Oxford University Press, Walton Street, Oxford OX2 6DP

Oxford New York
Athens Auckland Bangkok Bombay
Calcutta Cape Town Dar es Salaam Delhi
Florence Hong Kong Istanbul Karachi
Kuala Lumpur Madras Madrid Melbourne
Mexico City Nairobi Paris Singapore
Taipei Tokyo Toronto
and associated companies in
Berlin Ibadan

Oxford is a trade mark of Oxford University Press

Published in the United States
by Oxford University Press Inc., New York

British Library Cataloguing in Publication Data
Data available

Library of Congress Cataloging in Publication Data
Logue, James.
Projective probability / James Logue.
— (Oxford philosophical monographs)
Includes bibliographical references and index.
1. Probabilities. 2. Logic. 3. Realism. 4. Subjectivity.
I. Title. II. Series.
BC141.L64 1995 121'.63—dc20 94–36456
ISBN 0–19–823959–9

1 3 5 7 9 10 8 6 4 2

Typeset by Graphicraft Typesetters Ltd., Hong Kong
Printed in Great Britain
on acid-free paper by
Biddles Ltd., Guildford and King's Lynn

*This book is dedicated to
my mother and father,
Kathleen and Jack Logue,
with love and gratitude*

'nothing between human beings is one to three. In fact,' Sam the Gonoph says, 'I long ago come to the conclusion that all life is six to five against. And anyway,' he says, 'how can anybody let such odds as these get away from them? I think I will take a small nibble at this proposition.'

Damon Runyon:
'A Nice Price'

PREFACE

T H I S book presents, from a quasi-realist perspective, a univocal theory of probability which I shall call strong coherentist subjectivism (SCS). This theory advances:

- a *personalist claim* that the degrees of partial belief of (ideally rational) agents may be treated as interpretations of the measure 'probable' of the formal probability calculus;
- a *strongly subjectivist claim* that *only* such degrees of belief can be construed as probabilities and that the notion of objective chance is, if not conceptually impossible, at any rate redundant; and
- a *coherentist claim* that the single minimal constraint on beliefs of weak coherence (defined as immunity from Dutch books—betting arrangements which guarantee loss for the agent) is a sufficient, as well as necessary, condition for the rationality of those beliefs.

The mathematical basis and philosophical starting-point of the book is the work of Bruno de Finetti, which constitutes the most advanced and comprehensive mathematical articulation of the theory but which frequently substitutes for any philosophical underpinning a rather unconvincing and sketchily argued polemic. Consequently, I shall often be engaged in discussing issues which de Finetti glides over or even disdains to recognize as issues.

I shall be approaching these issues from the perspective which Simon Blackburn has christened 'quasi-realism'. Its contribution in this context is to allow one to see probabilities as projections of subjective evaluations which, provided that they conform to the standards of coherence, come legitimately to be expressed in apparently realist or objectivist language. Although I believe that SCS can serve as a complete interpretation of probability even without being linked to the wider quasi-realist outlook, I shall claim that that outlook provides a convincing rationale for the theory, helps to free it from anachronistic psychologism, and provides it with smoother solutions to long-standing problems. To mark this commitment, in later chapters I sometimes refer to my position as projectivist rather than subjectivist, indifferently referring to the theory either as SCS or as PP—a theory of *projective* probability.

To be a little more concrete about the task ahead, let me first give here a (necessarily crude) outline of subjectivism, taken from a recent introductory work, Weatherford [1982: 9–10]:

The principal views of this school are as follows:

(a) Probability is the degree of belief of a given person in a given proposition.

(b) Probabilities are best established by examining behaviour, especially betting behaviour.

(c) There are no objective probabilities—or at least this is a different and less important sense of 'probability'.

(d) An event has no unique probability. Each individual is logically free to set his own values.

(e) The probability beliefs of a rational individual must be consistent and governed by the Calculus of Probability.

Even in such stark outline, one can see both the appeal of subjectivism—as Kyburg [1978: 158] puts it, it appears 'maximally tolerant and minimally restrictive'—and the unease it generates: can one really build the whole of probability theory on such slight, and subjective, foundations? My thesis will be that one can.

Before that, though, the Introduction will argue that one needs to do so. To point up the need and scope for personalism, I compare it with the four main types of objectivist theory, asking of each three key questions: what are the objects of probability-judgement?; how are probabilities initially derived?; how do probabilities rationally change?— and I conclude that none of the objectivist competitors can adequately answer all these questions.

Chapter 1 will then deal with the statics of coherent belief. I begin by comparing various non-objectivist approaches to the nexus of partial beliefs, desires, and preferences, and then examine some of the problems involved in seeing bets as a measure of degree of belief. After presenting a definition of probability and some fundamental theorems, I then consider the connection between coherence and consistency and examine the role of Dutch books. I present an account of coherence as regulative (rather than simply normative or descriptive) and dismiss a number of attempts to sever the link between coherence and actual judgement.

Chapter 2 then turns to the dynamics of the theory, exploring belief-change and the mechanisms of convergence. I offer a relatively novel account of Bayesian conditionalization and attempt to show how along with the notion of exchangeability it leads to a refined theory of convergence of our judgements in which ultimate consensus, even when informed by frequency information, is yet explicable in purely subjectivist terms. I attempt to defuse the threat posed to this programme by the notion of weight of evidence with an analysis of this notion in terms of resiliency and higher-order probability.

Chapter 3 begins with a capsule comparison of SCS with other versions of personalism. It then considers whether the defects of the objectivist theories could be overcome by combining some or all of them in some schema of polyvocal senses of 'probable', and rejects a number of attempts at such a pluralist approach. It moves on to ask in what the subjectivity of SCS consists, rejecting a number of misguided views of this, exploring various possible views of SCS as anti-realism about probability, and arguing that the theory is best seen as a quasi-realism about probability.

Chapter 4 considers some applications of the projectivist theory. It repudiates any excessive Bayesian zeal which would claim that the theory disposes of the problems of induction and of scientific theory choice; it demonstrates the positive value of the theory in helping resolve various problems about explanation and conditionals and within judicial decision-making.

I conclude with a brief section making explicit the quasi-realist underpinning of the theory.

There are three important omissions from this discussion. First, I avoid dealing with many of the central problems of decision theory and utility: §1.1 explains why. Secondly, I avoid dealing with issues of inductive logic and confirmation theory: §4.1 explains why. Lastly, I do not consider whether the account is adequate for the quantum-mechanical context: this is an interesting issue, since many important questions of interpretation there centre upon the status of probabilities as physical quantities and the role of the observer—but more space is required than is available here to deal with this area properly. None of these topics will quite vanish from sight, but they will all be banished to the periphery —not because they are unimportant, but simply to clear centre stage for an exploration of the nature of probability itself.

It is also important to emphasize, even at this stage, a point which I have already mentioned: that the theory to be developed, while strongly subjectivist, will not be 'Bayesian'—at any rate, not in any of the stronger senses of that catch-all term. In a weak sense, any position on the foundations of probability which permits the wide or unrestricted use of Bayes's theorem may be described as Bayesian. Since, in general, those who object to such unrestricted use of the theorem do so because they argue that the initial probabilities used in the formula should be conceived of as objective and so would in many cases be unknown, acceptance of Bayesian modes of moving from prior to posterior probabilities usually accompanies tolerance for some personalist interpretation of probability. But in calling a theory 'Bayesian' one usually intends more than this. As Eells [1982: 4] puts it:

Bayesianism is usually characterized as the philosophical view that, for many philosophically important purposes, probability can usefully be interpreted subjectively, as an individual's 'rational degree of belief' . . . The subjective interpretation of probability is connected, however, in very important ways with a mathematically precise and intuitively plausible theory of rational decision, called the 'subjective expected utility maximization theory'.

The position which I shall adopt will involve no commitment to, or against, a utility-maximizing model of decision-making in general. Nor will it be Bayesian when it comes to decision-making within science: I shall reject what I see as the core thesis of scientific Bayesianism, expressed, for example, by Howson and Urbach [1989] in the claim that theory choice is essentially a matter of comparison of probability in the light of evidence.

It will also be apparent—I hope not too irritatingly—that the argument is constructed in a layered style: issues are raised in order to be dealt with partially, then returned to after further development of other parts of the theory. I present the theory in this way because it seems to me the most appropriate way to handle the range of criticisms which have been made of subjectivism, from the very crude to the sophisticated, and because much of the positive thesis involves advancing on several fronts at once. This will mean rather a lot of cross-reference and some, inevitable, repetition—which I try to keep to a minimum.

ACKNOWLEDGEMENTS

I HAVE been thinking about the philosophical foundations of probability for a considerable period of time. The text from which this book has emerged is that of a D.Phil. thesis of the University of Oxford submitted in 1989; but that thesis contained elements of an earlier B.Phil. thesis [1987], as well as traces of work done in London in the early eighties. Over that time, my views have progressed—or, at least, changed—from an initial scepticism about the possibility of any successful univocal theory of probability; through wild enthusiasm about the ability of a univocally subjectivist (Bayesian) theory to solve all the problems of probability, confirmation, induction, and scientific method; through sober realization that such enthusiasm could not be justified; to an understanding that confidence in the success of strong univocal subjectivism as a theory of probability can and should coexist with clear-headed awareness that it cannot alone function as a complete solution of those wider problems; and, finally, to recognition of the potential of a quasi-realist outlook for reconciling strong subjectivism with the apparently realist nature of much discourse about probabilities.

I have been helped, along the way, by discussion with, or written comment from, a number of people—almost all of whom, it should be said, would disagree more or less vehemently with the position I have finally adopted. They include Phillip Dawid, Michael Dummett, Ellery Eells, Donald Gillies, John Lucas, and Hugh Mellor. Three other figures (who have all, at different times, been supervisors of my graduate work) stand out, for me, as most influential, through their encouragement, suggestions and, often, disagreement, in enabling me to develop my ideas.

Andrew Wright set me on my way by first convincing me of the importance and merit of attempting to defend a strong subjectivist theory rather than falling back on an ill-motivated and woolly pluralism: a merit which depends not so much on the correctness of univocal subjectivism as on the inadequacy of existing defences of it. His patient support for my early efforts was utterly indispensable. Jonathan Cohen's influence on me has been profound, as will be apparent throughout the book. The powerfully original and independent account of induction and probability which he has developed over many years exhibits such internal cohesion, such wealth of knowledgeable detail, and such breadth of concern with scientific, forensic, and everyday reasoning as to demarcate, as no one else has demarcated, the problem-set for any theory of probability with pretensions to adequacy. At the same time as posing the most formidable of intellectual challenges to my views he has offered all-important personal support for my efforts to meet that challenge —though fully aware that such efforts must include attacks on his own theory. In the result, as it appears in this book, the dictum that in philosophy 'criticism is the sincerest form of compliment' for once has real force.

I first encountered Simon Blackburn's quasi-realism—which he has initiated, developed, and defended virtually single-handed—when my views on probability were already fairly fully formed: it shed an entirely new light on what I had been striving to do. Both by intellectual example and long-term personal encouragement he has shown me how an apparently narrowly confined enterprise can have ramifications for large areas of philosophy: I hope the outcome here of this new outlook on my original enterprise will demonstrate some of the force of the global quasi-realist approach and add to its persuasiveness.

CONTENTS

Introduction

If you look at ideas about probability and its application, it's always as though *a priori* and *a posteriori* were jumbled together, as if the same state of affairs could be discovered or corroborated by experience, whose existence was evident *a priori*. This of course shows that something's amiss in our ideas . . . (Wittgenstein [1975: §232])

THAT there is something very much amiss in most thinking about probability is one of the few propositions most probability theorists would agree on; even then, the proposition is likely to be treated indexically, the theorist's own account being taken to encompass the small minority of unconfused thoughts on the subject. Such partisanship is common enough in philosophy, but in this area it tends, less usually, to be accompanied by an urge to plunge into detail rather than fight out the central metaphysical and epistemological issues raised by the contending theories. A standard dialectical pattern is to offer a taxonomy of theories of probability—anything from a twofold classification (Carnap [1950]) to the elevenfold of Fine [1970]; defend in detail the applicability of one interpretation in all circumstances, or some (even all) in some specified varying circumstances; and criticize in detail the mistaken claims of other interpretations to be applicable in contexts where they are not. I would not disparage that pattern; something like it must surely provide the gist of any explanatory theory, and much of this work will conform to it. But it brings with it the danger of failing to address the larger issues of the force and motive of our detailed theory; I shall constantly, especially in later chapters, attempt to respect that danger. Wittgenstein [1975: §235] speaks of probability laws as 'the natural laws you see when you screw up your eyes': as in natural philosophy, so in philosophy, much can be gained at certain stages from deliberately blurring one's vision.

Much of this Introduction will be deliberately blurred (apart from whatever is accidentally blurred). My aim will be to demonstrate that there is room for argument in favour of a univocal personalist theory of probability—but I shall not say yet what form that theory might take.

This attempted demonstration will stem from asking three deliberately vague questions—what are probabilities, how are they known, and how can they rationally change?—of theories which will be, deliberately, crudely and incompletely presented and classified; it will ignore some well-known and powerful criticisms of the theories I reject and indicate criticisms of the one I advocate (ground which is well covered in Weatherford [1982] and Cohen [1989]); it will reject various forms of pluralist theory but conclude by indicating ways in which pluralism might be defended.

Mackie [1973: 155] has claimed that 'none of the . . . concepts [i.e. interpretations of probability] can be ruled out on grounds of internal health'. I disagree with that diagnosis for objectivist theories but I am not concerned here with attempting to write their death warrant. My hope is that indicating some of their internal problems will, apart from casting doubt on this cheerful diagnosis, show that whatever internal problems afflict personalist theories are at any rate very different from the objectivists'. Hence, if we are not convinced by the arguments for pluralism which I attack in Chapter 3 it becomes a worthwhile first move to look to personalism as potentially a less problematical approach to understanding probability.

I propose to consider five broad groups of theories:

- classical (C), which define probability as a ratio among equipossible alternatives;
- a priori (AP), which define probability as a measure of a logical relation between a proposition and evidence or as a range-function across possible worlds or state-descriptions in a logical language;
- relative frequency (RF), which define probability as the (limiting) relative frequency of some characteristic in some (infinite) sequence;
- propensity (PROP), which define probability as a physical property of some thing, or some experimental set-up, or a disposition to display such a property, or the capacity to evince such dispositions;
- personalist (PER), which define probability as the degree of belief an actual or ideally rational individual might or should have about some outcome.

At least two important groups of theories are omitted from this classification. First, it ignores multi-valued logic interpretations (stemming from Reichenbach [1949]); nothing useful can be said about them until the relationship between probability and provability has been explored. Secondly, it ignores 'guarded assertion' views (stemming from Toulmin

[1950]); I believe that that view is properly regarded as an account of the expression-conditions of probability-judgements which emerges naturally from the quasi-realist analysis which I shall undertake in Chapter 3.

For each of these groups of theories I ask:

(*a*) What is (rightly said to be) probable?
(*b*) How can probabilities be initially determined?
(*c*) How can probabilities change in the light of experience?

(a) *The Content of Probability-Ascriptions*

In its pure form, C would restrict probability to situations on which an equipossibility model could be imposed as a plausible consequence of symmetry or perceptions of symmetry. We could not then legitimately speak of the probability of such commonplace events as rolling a five with a biased die or of the Labour candidate winning a by-election against opposition from Conservative, National Front, and Monster Raving Loony candidates; we could deal neither with the probability of a unique event nor with actuarial probabilities. Such restrictions are altogether implausible; indeed, the early 'classical theorists' did not attempt to hold that line, Laplace and Daniel Bernoulli adopting a frequency definition for probable mortality and Jacques Bernoulli explicitly denying that C could encompass actuarial cases which, he insisted, none the less involve genuine probability.

AP theories are superficially much more open, treating indifferently both singular and repetitive events. But we must recall that this tolerance is severely circumscribed by the AP view that what is perceived in any probability-appraisal is not any property of the objects appraised but rather a relation between propositions about the object or event. For it is one thing to insist that probabilities must be assessed in the light of evidence, quite another to *define* them as the measure of the relation between a proposition and evidence. We must, then, take the statement 'it will probably rain today' as elliptical for 'it will probably rain today given falling barometers etc.'; the statement cannot then be taken to be about an impending change in the weather—rather it is about the strength of reason provided by some evidence (barometric, etc.) to expect rain today. This is objectionably counterintuitive, at best.

RF theories find it necessary to distinguish between repeatable and singular events, and, in general, to exclude the latter as objects of probability-appraisal. This has often, and rightly, been seen as one of the respects in which RF theories show up as most restrictive. Von Mises [1957] recognizes this point, but tries to draw its sting by comparing his programme for refining the pre-scientific concept of probability with abstraction from the everyday notion of work to the rigorous concept 'work' in mechanics. He claims that, just as the latter case abstracts all that is useful for science to create a new concept without denying a (non-scientific) meaning to the loose everyday usage, so by limiting use of the scientific concept of probability to collectives of repeatable events, one preserves all that can be scientifically useful of the concept.

What does RF theory exclude in excluding unique events? First, it declines to deal with—as of no concern to science—such judgements as 'it is probable that America's bombing Beijing would lead to war' and 'it is probable that Augustus wept over Varus' loss of the legions'. Excluding the first type invites the response that we are excluding from our theory just those cases which are of most importance for action: probability will cease to be a guide to life. So, if I want to know how to act on discovering the American intention to bomb Beijing, I may expect psychology or sociology to help me assess various factors relevant to my decision, but no RF probability science could help me combine those factors. When the historian claims that Augustus probably did weep, must I suppose that a different kind of criterion applies to his or her judgements on the veracity of Livy or the authenticity of extant manuscripts, from that which applies to his or her judgement as to combining these factors into a single judgement? Or must I dismiss all his or her judgements as 'non-scientific'? Secondly, RF theories must exclude as objects of probability-appraisal at least those theories of universal scope: it seems impossible to create a collective of Special Relativity theories and measure the frequency with which they work. So, a theory which is indisputably scientific cannot scientifically be claimed to be probable, or more probable than another, or less probable now a particular fact has been observed. And there is no doubt that Von Mises's position commits him to this 'scientifically'—not just 'testably' or 'as a result of actual observation'.

More strongly, is the distinction between singular and repeatable events even a tenable one? Both frequentists and some of their critics seem to have accepted a rather simplistic distinction. Unique or singular events, it is said, are those which necessarily cannot or in fact do not form part of a collective: Caesar may have crossed the Rubicon, but we cannot, even in principle, assess the probability that he did by lining up ten thousand Caesars faced with the same choice or exhuming Caesar to have him make the choice ten thousand times with amnesia between tries. On the other hand, we can readily repeat the event of tossing a coin to form a collective in either of these ways: I may toss ten thousand similar coins or I may toss this coin ten thousand times under similar conditions. Of course these tosses are not identical—they occupy different segments of time and/or space. A collective must consist of a number of events, dissimilar in various ways, but similar in some important respects. Now consider this experiment. A historian takes ten thousand men, selected to be ambitious for power, and equips each of them with an army whose ostensible mission is to defend a small state; each is asked to consider using the army to take over rule and put in a position where his decision is irrevocable and public. (S)he now considers the collective consisting of these men and Caesar two millennia earlier, and by measuring the relative frequency of one aspect of behaviour arrives at the probability that Caesar, considered as a member of this collective, did cross the Rubicon. How does this differ—other than pragmatically—from the coin-tossing case? Or again, compare the similar process where an actuary attempts to assess the probability that I will die in the next year, treating me as a member of a ten-thousand-strong collective of untubercular European men. I am a unique individual; my death would be a unique event; but a collective of similar individuals and events can be formed. If the actuary is more likely to be right than the historian, that can only be because I can be described to him or her so as to fall into a collective small enough so that enough relevant factors will remain constant but wide enough for statistical information to be gathered—whereas we know so much about Caesar's behaviour that it is difficult to form a large enough collective and so little about his motives that we do not know which factors are relevant in selecting our similar men. But in broad terms, both cases are methodologically similar and both resemble our coin-tossing paradigm.

To sum up: since a collective is a set of events with some differing properties, the notion of repeatability must be weakened to that of selection of members with no relevant differences. If RF permitted a theory of relevance, we might be able to use it to distinguish the easily generalizable from the less easily generalizable among events. Without such a theory there is, perhaps, a spectrum of ease and accuracy of collective-forming, but no methodological differences along it.

Turning now to propensity theories (PROP): early versions of such theories were unclear whether to try to accommodate the concept of the probability of a singular event, and Popper [1959: 260] conceded that his interpretation 'may be presented as retaining the view that probabilities are conjectured frequencies in long . . . sequences' and [1957: 67] that probability is 'the propensity of the experimental arrangement to give rise to certain characteristic frequencies when the experiment is often repeated'. Suppose the experiment cannot be, or in fact never will be, repeated? One possible solution to this difficulty which Popper has increasingly emphasized is to insist that propensities are dispositional properties of the conditions generating results rather than properties of the results, of which we might have a long sequence, or one, or even none. To put the problem in a form which perplexed Peirce [1932: iii. 281]:

If a man had to choose between drawing a card from a pack containing twenty-five red cards and one black one, or one containing twenty-five black cards and one red one, and if drawing a red card were destined to transport him to eternal felicity . . . and a black one . . . to eternal woe, it would be folly to deny that he ought to prefer that pack containing the larger proportion of red cards, although from the nature of the risk it could not be repeated.

Popper's solution would allow us to prefer the red-dominated pack while continuing to view probabilities as physical properties, not ratios, since the properties involved attach to the constitution of the deck, regardless of how many trials of the experiment are possible. This manœuvre will, however, lead to difficulties about change in probabilities—see (c) following.

PER theories have the substantial merit of experiencing no strain in dealing with probabilities of single events, events which the RF theorist would call 'repetitive' and collections of such events (though, as will be seen, they too find difficulty in assigning probabilities to scientific generalizations). But there are a number of challenges which they must meet.

1. Are probabilities, thus interpreted, simply in the mind rather than in the world?—or at least made rather than discovered?
2. Are probabilities just any beliefs or only ideally rational beliefs?— if the latter, are PER theories normative?
3. Is belief really a more primitive notion than probability?—could we not claim (as in Day [1961]) that my believing something partially can only be understood as my accepting that it is probable to some degree that it will come about?

(b) *Sources of Initial Probabilities*

C theories rely on some version of a principle of indifference to provide initial values for a probability ratio. Notoriously, this principle, if un-qualified, leads to many paradoxes, especially if applied to continuous distributions: the worst problem here emerges from generalizing Von Kries's relative density example (see Blackburn [1973: 122–6]), which produces irresoluble difficulties if the principle is to lead to equiprob-ability. It is arguable that we should be reluctant to abandon entirely a principle of such powerful intuitive appeal. Blackburn suggests that the principle might have application to the rationality of treating alterna-tives with equal degree of confidence as reasons even where one could not reasonably believe they were equally probable; Benenson [1984] advocates taking the principle as a means of fixing identity conditions for probability-judgements in terms of relations of evidential support (construed on an AP model). But even if the principle has some such role it will not, independently of some other conception of probability, fix initial probability-values. For, as Hume pointed out in the *Treatise* (I. iii. 11), 'Where nothing limits the chances, every notion that the most extravagant fancy can form is upon a footing of equality.'

For AP theories, since probability is a logical relation we arrive at probabilities by discovery of that relation, either by direct intuition (as in Keynes, for some probabilities at least) or by calculation (as in Carnap). To make the former at all plausible, we must suppose that not all probabilities can be assigned numerical values, and indeed that any two arbitrary events may not even have directly comparable probabili-ties. But to admit this is to admit that, in many cases, one is entitled to arrive at some intuition which can be termed a 'probability' but which is not a measure-function conforming to the probability calculus. The latter option involves utilizing concepts of range (as suggested in the *Tractatus*, 5.15) or of possible worlds (e.g. Bigelow [1976]) or of con-firmation-functions (Carnap [1950]) to calculate a ratio of the state-descriptions where an event occurs to the total of state-descriptions possible in any finite language. But then it would seem that probabili-ties are dependent on the resources of our language or our choice of confirmation-function; and, as is clear from Putnam [1963], we cannot even in principle establish one best confirmation-function for all pos-sible worlds, or even for our own world.

For RF theories, this issue at least poses no insuperable problems. Either one can take the Von Mises view that probability theory is simply a means of establishing methods for constructing sequences or collectives or, what may come to the same thing, follow Reichenbach in claiming that all probabilities are, and can only be, a posteriori determinations of relative frequencies. Then there will be no point in distinguishing initial probabilities from probabilities modified by experience since all probabilities are wholly defined by sequences of experiences of outcomes.

For PROP theories, we presumably arrive at a probability-judgement from observation of a particular experimental set-up, or by forming a conjecture as to a propensity's magnitude which is then tested using a particular experimental set-up. The question then arises of how to connect descriptive attributes of a set-up and causal stories about its subsequent states with numerical values for probabilities. PROP theories have no distinctive numerical implications as long as they speak the language of causality, not of ratio.

PER theories offer the most radical answer to the question of whence to derive initial probabilities: any such derivation you can make coherent with your other beliefs is legitimate—you may have an infinite number of such possibilities, and another individual may have a different infinite range. Numerical values can be attached to your judgements by eliciting from you, perhaps with some idealizations, the odds you would accept on bets about each outcome.

These theories then face four main questions.

1. Does such alarming permissiveness mean that no one coherent distribution of beliefs is ever more reasonable or more justified than any other?
2. Why and how does actual widespread agreement of judgements come about if any extreme but internally consistent position is possible?
3. Does using betting-measures to provide numerical values for degrees of belief mean that utility is a more primitive notion than probability?
4. Are there not some circumstances where settling bets is impossible, yet we do, and it seems right that we should, form probability-judgements?

(c) *Learning from Experience*

C theories are quite helpless to account for the modification of probabilities by experience. Presumably, if I make an initial judgement, based on equipossibility of the outcomes represented by the six faces, that the probability of rolling a six with this die is $\frac{1}{6}$, and then find it tends to come down six as often as not, I might change my mind and hold that six and not-six were equipossible. Then again, I might not. Since there is nothing in C to account for the origins of my perceptions of symmetry no mechanism exists for those perceptions to be corrected. And, obviously, it will not do to assume (as Huygens, for instance, did) that all competent judges armed with the same body of evidence will always assess probabilities at the same value.

AP theories, equally clearly, cannot readily account for observations altering our probability-judgements: for if what is involved in such judgements is a recognition of a relation between propositions, what happens in the world would seem to have no direct bearing on that logical property. An AP theorist might respond that, given that all probabilities are evidence-relative (being logical relations between a proposition and the totality of available evidence), and that as events occur in the world this totality enlarges, although the initial probabilities remain as relations to the original evidence, we will naturally be more interested in the probability of the event relative to the whole evidence that there is now. But this answer just asserts that it is possible to distinguish one probability-judgement from another in virtue of their evidential base: it goes no distance towards showing how *we*, who do not at any point have access to the totality of the world's evidence, ought to react on acquiring some new information. For, if the fact we encounter was part of, or compatible with, the initial set of state-descriptions which (unknown to us) determined the logical relation, we ought not to change our probability-assignments; if it was not, we should. Lacking the reference point of an individual's known state of information, we are at a loss how to proceed. As Ayer [1957: 73] put it:

if we are presented only with a stock of necessary facts to the effect that certain statements, or groups of statements, bear logical relations to each other . . . I do not see what reason there could be for differentiating between the items of this stock as bases for action.

RF theories fail for a very simple reason: they define probability in terms of relative frequency in infinite sequences, and we so seldom live through infinite sequences; so any finite set of observations will be compatible with any frequency over the infinite sequence. The only possible defence here would lie in precisifying the notion of a limit in an infinite sequence which could be apprehended in a long, but finite, run; but, while it is known how to define limits for sequences which obey some rule of construction, it has proved quite impossible to do so for random sequences—the whole idea of randomness tells against it: see, for example, Martin-Löf [1969] and Howson and Urbach [1989] for a survey of the difficulties.

PROP theories, as we saw in (*a*), can only cope with single-case judgements by taking probability to be a property of the experimental set-up, not of the results of a sequence of trials. But if that is the case, it is difficult to see how our judgements of probability can be affected by the results we experience except in so far as these can be mediated through some scientific theory showing that the experimental set-up necessitates the results: it is precisely because we lack such a theory that, normally, we turn to probabilities at all.

PER theories have the advantage of being able to make freer use of Bayes's theorem here to move from prior to posterior probabilities than objectivists can normally feel happy with. So, they can hope to explain how experience changes probabilities purely in terms of the modulation of simple degrees of belief into the degrees of belief which had been assigned conditional on E, once E is known to have occurred or thought probably to have occurred (once it becomes evidence). Just what is involved in such conditionalization, and what constraints of rationality must be imposed on it, forms an eighth question our theory must tackle.

1

Coherent Belief

THIS chapter concerns itself with the statics of the version of personalism which I am advocating: strong coherentist subjectivism (SCS). Questions of dynamics—about changes in belief-states—will be deferred until Chapter 2.

§1.1. asks whether partial belief, or belief plus desire, or preference is the fundamental concept on which a theory of probability should be built.

§1.2. considers the problems of measuring partial belief by dispositions to bet and enquires whether SCS need be committed to an operationalist view of measurement.

§1.3. relying directly on de Finetti's work, offers a definition of probability in terms of coherence and states some fundamental theorems which will be important to later chapters.

§1.4. examines the connections between the coherence constraint, deductive consistency, and the formal probability calculus and explores five areas of difficulty for the project of basing probability on coherence.

§1.5. asks whether SCS is a descriptive or normative theory, meets criticisms stemming from the work of the psychologists Kahneman and Tversky, and concludes by presenting the theory as an idealized descriptive theory of actual probabilistic competence which thereby has regulative force.

1.1. BELIEFS AND DESIRES

(a) *Uncertainty*

Suppose that I am about to toss a coin: do I believe that it will come down heads? It seems implausible to insist on a 'Yes' or 'No' answer to such questions. We want to be able to reply: 'I believe it to a certain degree' or 'I partly believe it and partly believe its negation' or, perhaps,

'I partly believe it and partly do not believe it'. To account for our attitudes in such contexts we need some concept of gradable partial belief.

Suppose that I am about to toss a coin: will it come down heads? When we cannot answer 'Yes' or 'No' to such a question, we often want to be able to say: 'It is to some degree probable that it will land heads.' To account for the outcomes in such contexts we need some concept of probability.

The fundamental question then arises as to the relation between partial belief and probability: depending on the answer given by a theory of probability, we might characterize such a theory as subjectivistic or objectivistic. Pursuing the first, rather crude, approximation followed in the Introduction, we might say that subjectivistic theories take partial belief as primary, defining probability as its measure, while objectivistic theories take probability as primary, treating partial beliefs as no more than estimates of probabilities or denying that the concept of partial belief is of any use. This distinction is useful, but it is important to be alive to its limitations. Objectivists have often shown a hostility to the concept of partial belief which is quite unnecessary and damaging to their positions. Among frequentists, Von Mises's [1957] dismissal of the concept and Reichenbach's [1949] repudiation of any larger role for partial belief than in personal estimation of relative frequencies lead to enormous difficulties in connecting their theories to action. Among logical range theorists, either treating partial belief (as Keynes [1921] does) as introspective detection of logical relations, or denying (as Kneale [1949] does) any need for a conception of degree of confidence in A over and above the degree to which the evidence probabilifies A, founders on the fact, to which Kneale himself [1949: 20] draws attention, that 'no analysis of the probability relation can be accepted as adequate . . . unless it enables us to understand why it is rational to take as a basis for action a proposition which stands in that relation to the evidence at our disposal'. Nor, on the other hand, is it necessary for theories which make partial belief central to be subjectivistic: rather, they may adopt an objectivistic variety of personalism, as in Mellor's [1971] system, where partial beliefs are what is expressed by probability-judgements, but what makes such judgements reasonable or unreasonable is a corresponding objective propensity. Other versions of personalism which attempt to make room for a concept of objective chance will be discussed in Chapter 3.

By the same token, although in the account of probability which I am urging—a strong coherentist subjectivism—we are committed to the claim that 'only subjective probabilities exist, i.e. the degree of belief in the occurrence of an event attributed at a given instant and with a given degree of information' (de Finetti [1974: i. 3]), this commitment, as I shall argue later, does not have the alarming consequences which might be anticipated. But certainly partial belief is central to it.

It would be futile at this stage to attempt to characterize partial belief much more closely than simply taking 'partial' to indicate neither full endorsement nor outright rejection: the project of developing a strongly subjectivist theory of probability will provide an explication of the concept. But it is, even this early, worth noting some of the places where the theory will have to take a stand and some where it may be able to avoid commitment. First, hostility towards objectivism about probability need not lure us into identifying partial beliefs with introspectible feelings of some kind. De Finetti is much too ready to do this: presenting himself as following Hume in this respect, he identifies partial belief with probability and writes of 'one's own sensations of probability' [1974: i. 84]. But this is extremely problematical; for, as Ramsey [1931: 169] realized, 'the beliefs which we hold most strongly are often accompanied by practically no feeling at all; no one feels strongly about things he takes for granted'. And this is a problem which does need to be tackled immediately, as I shall do below. Secondly, whether we reject such 'Humean' notions or are merely agnostic about them, if we are not to rely on them we must give some alternative structure to the notion of partial belief. There are several ways in which one could do this in terms of dispositions to act: (*b*) and (*c*) below and §1.4 will raise the question which of these is to be preferred. Thirdly, our theory needs to be unambiguous as to whether it is to deal with the actual beliefs of actual people, with all their imperfections on their heads, or some sort of idealized belief-states of ideally percipient agents committed to believing all and only the logical consequences of their beliefs: §1.5 will address this issue. And, lastly, the question will arise—though I shall not tackle it until the Conclusion—whether an important issue about the content of probability-judgements is concealed by an act–outcome ambiguity in the notion of 'partial belief'.

Returning to the first question, this issue has recently been given added urgency by Jonathan Cohen's [1992] arguments that there is a distinction of fundamental importance (over a much wider range than subjective probability, but crucial here) to be drawn between belief and acceptance. Belief and acceptance, as Cohen has characterized them, differ in at least five main ways:

1. belief is an inner disposition to feel, acceptance a commitment to adopt;
2. belief is involuntary, acceptance voluntary;
3. belief can come in degrees, acceptance is all-or-nothing;
4. believing *p* does not entail belief in all the deductive consequences of *p*, accepting *p* does entail accepting them;
5. believing *p* is related to *p*'s truth in some way in which accepting *p* is not.

One obvious immediate question, which I shall not pursue here, is whether characterizing belief in this way shrinks the notion to vanishing-point—that is, quite independently of any hostility we might have towards 'folk psychology' or epistemology in general, do we have any use for a concept of inner, involuntary dispositions to feel it true that . . . ?

A question which must be discussed is why acceptance should be taken to be an all-or-nothing matter; or, if one *does* prefer to reserve the term 'accept' for all-or-nothing choices, why should there not then be important acts of judgement which are gradable, voluntary, commitments? Cohen has presented acceptance as distinct from assent: it is having a policy of 'going along with' a proposition in one's mind. Cannot I, in an everyday sense, go along with a proposition wholeheartedly, or half-heartedly, or three-quarters-heartedly? Granted, if one operates with some such concept of gradable commitment to *p* one cannot then evade the traditional, and formidable, problems surrounding setting a threshold of that gradable commitment above which it is reasonable to adopt or act on *p*—but these are problems which must be faced in one form or another anyway.

Moreover, it is unclear just what position Cohen is adopting as to whether there are subjective epistemic commitments which resemble belief in some of the five ways I have mentioned, acceptance in others. A relatively weak thesis would be that the gradable state of belief-as-feeling cannot in general suitably be connected to an eventual, non-gradable, act of acceptance, though there might well be other gradable judgements suitable for such purposes. If there is a stronger thesis, it presumably must run something like this: there are objective degrees of acceptance-worthiness, and subjective degrees of willingness or inclination to accept a proposition, but no gradable subjective commitment to the proposition itself other than more-or-less intense belief-feelings. This seems very implausible. What argument would support a claim that level of gradation could vary only with the content of the proposition, and not attach to some suitably defined form of willingness to accept the proposition which is a gradable function of an agent's act of judgement concerning the proposition? I cannot see how one can reasonably rule out the very concept of guarded commitment by insisting that, belief-feelings apart, there can only be non-gradable acceptance of a gradable proposition or of something gradable about the proposition: how could one then explain when and how such acceptance is justified and when not?

Which version of the thesis is being advanced matters greatly to the subjectivist. On the weaker interpretation, the subjectivist is just being urged not to identify probability with the intensity of feelings of partial belief. That, I am sure, is good advice. As I shall argue later, subjectivism cannot get off the ground without some concept of coherence of probability-judgements; if such judgements were involuntary, coherence would be impossible to strive for, if subject to no deductive closure condition, too easily accidentally found. But this is something which subjectivists have increasingly come to recognize. Ramsey and de Finetti, I have conceded, certainly did identify probability with what they called 'partial belief'. But even then, Ramsey made it clear that he did not intend by that phrase intensity of feeling, that he was interested only in 'belief *qua* basis for action', that our degree of belief that *p* is best seen as 'the extent to which we are willing to act on it': even then, he had second thoughts (in his note 'Probability and Partial Belief') about his over-psychologistic presentation of his theory. De Finetti [1972: 189], despite his talk of 'one's own sensations of probability' almost always took a strong operationalist stance on the reducibility of partial belief to betting-measures, and came to see the judgement of probability as a task which could be performed more or less well, a matter of 'deciding on and adopting some point of view as a basis for previsions and related decisions'. The direction of evolution of subjectivist theory this century has been steadily away from interpreting probabilities as belief-feelings towards taking them to be behavioural dispositions—towards a theory centring on evaluations in which reference to feelings is, at best, inessential.

So, the weaker thesis poses no problems, if it offers no special support, for subjectivism. For it seems to come down to the claim that, while acceptance is all-or-nothing, both acceptance-worthiness (a property of a proposition) and willingness to accept (a property of an agent) are gradable and distinct from belief, and have a role in determining acceptance. The objectivist can say that objective acceptance-worthiness renders certain degrees of willingness to accept more reasonable than others; the subjectivist, that acceptance-worthiness is reducible to, or a projection of, willingness to accept.

But if what is intended is the stronger claim that gradation attaches only to the object or content, not the act, of judgement, that will be much harder to make compatible with a subjectivist theory. Such a claim will force, or at the very least strongly impel, us towards construing probability extensionally, rather than as constituted by a real or ideal agent's partial commitments—and that route leads to some form or other of objectivism about probability. So the subjectivist must, I think, resist any such strong thesis. We will need to show how, starting from an account of a graduated commitment to p, we can emerge with the same results we would derive from treating that gradation as attaching to the object—though without the objectivist's problems about evaluating it. This is a large part of the task ahead.

(b) *Beliefs, Desires, and Preferences*

Let us return to the coin waiting to be tossed. I may, of course, have other attitudes towards it than belief, full or partial, in (or acceptance of) some outcome: I may prefer or desire some outcome over others. We may wonder whether such an attitude should be represented as an attitude towards objects, or towards events, or towards propositions, or towards sentences. With this question, meanings enter the picture, and along with them the question whether an integrated theory of meaning and action is necessary to making sense of our behaviour under uncertainty. Even if we need not incorporate an account of partial belief in an account of meaning, is it necessary to give a unified account of preference encompassing both beliefs and desires?

There are, at the least, clear enough parallels between some central tasks within decision theory and within some visions of the role of a theory of meaning. One vivid way of presenting the options available diagrammatically stems from Peacocke [1979] (see Fig. 1.).

The concepts at the tails of each arrow have as resultant those at the heads. Resolving the head-concepts into those at the tails is achieved (or intended to be achieved) in the upper instance by a Davidsonian radical interpretation procedure, in the lower by a subjective decision theory or probability interpretation.

Five options suggest themselves for dealing with this network of concepts.

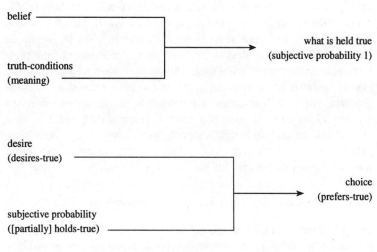

FIG. 1

First, we might, as Davidson has long urged and in part [1985] attempted, construct an integrated theory of meaning and decision in
which the upper arrow is seen as feeding into the bottom fork of the
lower arrow. On this model, our starting-point is preferences among
sentences (not propositions); we resolve these preferences into desires
and subjective probabilities; taking the subset of the latter consisting of
those sentences assigned probability 1, we solve these for beliefs and
meanings. Secondly, we might, as Blackburn suggests [1980*a* and 1984]
offer a unified account of both cases not by fitting one into the other
but by seeing them as parallel attempted reductions within the enterprise of 'quasi-realism'—the attempt to explain the realistic-sounding
nature of our talk of evaluations given that evaluative properties are
(merely) projections of our own sentiments.

Less ambitiously, we might leave the problem of meaning aside and
deal only with the lower nexus of concepts. So, thirdly, we might,
starting from what Savage [1954: 60] calls 'the fundamental empirical
situation of decision under uncertainty' decline to take either desires or
beliefs as primitive, instead attempting to explicate them simultaneously in terms of preferences among actions. The classic developments
of this approach are those of Savage [1954] and Jeffrey [1983]. Or,
fourthly, we might, in effect, resolve from choice to desires by holding
subjective probabilities constant while solving for desires. This is the

enterprise attempted by Von Neumann and Morgenstern in their [1947] decision theory: degree of belief is a primitive uninterpreted term which figures in the construction of measures of desirability. Finally, we might instead hold degrees of desirability constant while resolving from preferences to subjective probabilities. This is equivalent to what Ramsey [1931]—who is often mistakenly thought to have offered a theory of the third type—was attempting in constructing an account of subjective probability initially for 'ethically neutral' propositions; and it is what de Finetti is attempting in defining probability in terms of price or loss–gain on an agreed scale of utility taken as primitive and unexplicated until the theory of probability has been established.

(c) *A de Finettian Approach and its Commitments*

This treatment follows through the last of the options outlined above: consequently, although accepting that the interpretation of probability in terms of partial belief is importantly connected with a Bayesian theory of decision, I take it as possible, and of fundamental value, to establish independently of such theories the credentials of strong subjectivism as an account of probability. This position I take, essentially, to be that adopted in his seminal writings by de Finetti (although late in life he showed some leanings towards an approach akin to that of Savage).

It is important to note that there is a radical difference between the relation of my approach to the first two other options and its relation to the third or fourth—a difference which the double arrow illustration, I hope, brings out. We are not committed by taking up a de Finettian stance to embracing or rejecting the wider perspectives of the first two options on the larger context in which a theory of probability operates, the uses to which it may be put, or the programme it may be taken as exemplifying. Neither approach is undermined by taking the stance that it is important to establish subjective foundations for probability without reliance upon either a particular account of meaning or a particular position on realism. Consequently, although much of the later part of this thesis will be devoted to arguing for a quasi-realist view of subjectivism; although I claim that the theory is best understood from a quasi-realist perspective and that its successes add weight to the general quasi-realist case; still, the theory *need* not stand or fall with the quasi-realist interpretation of it advanced here. I shall be arguing that if we start by treating probability-ascriptions as non-truth-conditional stances—

commitments to loss–gain—then certain minimal rationality constraints on those commitments will render them indistinguishable from commitments to propositions with truth-conditions. But someone who is not convinced by the argument or who repudiates the very possibility of a quasi-realist enterprise can still accept the strong subjectivist theory as the best account of probability available.

We are, however, committed to rejecting the third and fourth options, and we need to ask why our own is to be preferred to them.

For Von Neumann and Morgenstern, it was important to construct an account of both probability and utility as a basis for the theory of games. A de Finettian theory would equally have served that purpose; and, it seems to me, there is a clear-cut reason for preferring a theory which takes utilities as primitive, constructing measures of probability, to one which does the reverse. The reason is, quite simply, that we are not concerned to develop a unique scale of utility whereas we do aim to develop a unique scale of probability-measurement (though of course that allows for a variety of methods of assessment of the measurements). If we adopt a de Finettian approach we need only assume that there is *some* scale of utility relative to which loss and gain may be defined, in order to set in motion our definition of probability. One way to proceed, which Ramsey saw, is to start from the subset of propositions comprising arbitrarily chosen ethically neutral propositions (i.e. propositions such that their truth or falsity is a matter of indifference to us); then we need only identify two states of affairs α and β such that we prefer α to β to enable us to define degree of belief $\frac{1}{2}$ in the ethically neutral propositions as a lack of preference between 'α if p, β if not p' and 'β if p, α if not p' for all p in our set (which may have only one member). We can then devise a measure of our degree of belief in any other proposition q, ethically neutral or not, by ascertaining our preference, or lack of it, between 'α if q, β if not q' and 'α if p, β if not p'. But to set in motion the Von Neumann–Morgenstern theory we need to rely on the unique, unexplicated concept of subjective probability as a transitive and additive normed measure. We ought to adopt the account of lesser commitment.

Turning now to the line taken by Jeffrey and Savage of simultaneously explicating desires and beliefs in terms of preferences, this seems to involve us in three difficulties which the de Finettian line avoids.

First, attributing probabilities and desirabilities to the same entities involves a reformist reinterpretation of certain ordinary ways of talking:

Jeffrey disclaims any such aim [1965: 60] but concedes [1965: 68] that his account must 'rule out certain natural kinds of talk'. While we normally speak of desiring objects, we do not, except in very atypical cases, speak of believing objects (believing in objects is a different matter). To achieve the unity his approach demands, Jeffrey stipulates that talk of desiring objects should be reinterpreted as desiring the truth of propositions, thus coming into line with belief. This is a risky programme: construal of desire as a propositional attitude is surely less plausible than so construing belief, whatever the difficulties in the latter case—we normally want to attribute desires to agents to whom we might hesitate to attribute beliefs.

Secondly, if desires and partial beliefs are to be explicated in the same way, we are forced to avoid attributing desirability to propositions known to be true. So, desirability is incompatible with known possession. This is certainly less natural than the de Finettian approach where, since I am to assess probabilities in terms of betting-quotients, it must be supposed that utility attaches to some objects which I know I possess (else how could I bet?).

Thirdly, treating beliefs and desires in a unitary manner tends to lead, without much choice in the matter, towards the unified accounts of meaning and action sketched in the first two options. If beliefs are, like desires, aspects of behaviour which can be attributed to non-language-users (animals like Jeffrey's cat, for example) the objects of belief cannot be sentences and must, presumably, be propositions; then, for us language-users just as for the cat, our beliefs are manifest not in our assent to sentences but in our holding propositions true. And how are we to account for what it is to hold a proposition true other than through an account of meaning which allows us to arrive at meanings out of aspects of behaviour (including acts of verbal assent)? This may seem less of a problem to those committed to a generally Davidsonian line; but surely, even if we were so committed, it would be preferable that some of our accounts of the world—our theory of probability, for instance—should not hang on the correctness of our account of meaning.

1.2. MEASURING DEGREE OF BELIEF

(a) *Bets*

The time-honoured way to find out how seriously a man is committed to what he says is to invite him to 'put his money where his mouth is'.

Although there may be an anticipatory hint to this effect in Kant (*Critique of Pure Reason* A825/B853), Ramsey [1931] and, independently, de Finetti [1937] seem to have been the first to suggest that betting behaviour might provide a suitable measure of the strength of partial beliefs, enabling the quantification of those beliefs so that they can be treated as interpretations of the uninterpreted term 'probability' of a formal probability calculus. With some variations, not important here, such calculi define probability-measures as functions p on a set S which attach to each subset A of S a real number $p(A)$ satisfying 3 conditions:

(i) $p(A) \geq 0$ for every A
(ii) If $A \cap B = 0$, $p(A \cup B) = p(A) + p(B)$
(iii) $p(S) = 1$

(These conditions, with some others, formed the axiom-set for the first rigorous set-theoretical formalization of probability, by Kolmogorov [1933].)

If p is interpreted as a function whose values are betting-quotients—that is, the values $p/p + q$ derived from betting-odds of p : q—this interpreted measure will obviously conform to these conditions.

It is important to note (though detailed treatment of these issues will be deferred until later) that the proposal to characterize partial belief by betting-quotients presumes that partial beliefs are comparable, may be precisely evaluated, and share the complementational character of betting at odds. All these assumptions are contestable.

First, it might be argued that even if some partial beliefs are comparable it need not follow that any partial belief is comparable as to intensity with any other belief: degree of belief might be non-transitive so that A being comparable to B and B to C does not entail A being comparable to C.

Secondly, Kyburg [1968 and 1974] has argued that the assumption that degree of belief can be evaluated to any required level of accuracy—implicit in a betting-quotient measure—is implausible in that it leads to licensing very strong beliefs about matters of fact on purely a priori grounds. He claims that degree of belief should properly be measured by an interval, not a real number. I consider this claim in Chapter 3.

Thirdly, it might be argued that many important cases of partial belief which legitimately count as probability-judgements are cases where the degree assigned to the belief and its negation are non-complementary.

But betting-behaviour, as we normally understand it, presupposes that bets are complementary. I consider this argument in §2.4.

Granting these assumptions for the moment, it is clear that betting-quotients can act as measures of subjective probability, in at least some circumstances, provided that we treat partial belief as dispositional and distinct from conscious assent and permit a modicum of idealization in our betting-measures from the behaviour of actual gamblers. If we wish, as we surely do, to be able to say that any uncertain event possesses a probability of occurrence, and we wish as subjectivists to identify that probability with partial belief, then we must treat such beliefs as dispositional or be forced to the wild claim that any individual at any moment is possessed of beliefs of a precise intensity about all events about which (s)he is uncertain. For the same reason, partial belief cannot be identified with conscious assent; independently, as Mellor [1980] points out, analysis of unselfconscious action and of cases of self-deception necessitates distinguishing the two. Finally, if the betting-quotients we deal with are to conform to the axioms of the calculus, it is clear that they must represent idealized versions of the odds taken and given in actual bets: stake size and direction of bet must be taken as irrelevant and a linear scale of utility presumed. Although Mellor [1971: 37] has provided strong arguments for supposing that a subjectivist may plausibly and innocuously effect the required idealization simply by taking as a measure of a man's partial belief 'the odds he will determine if compelled to bet where his opponent subsequently decides both the stake size and the direction of the bet', critics of subjectivism (e.g. Ryder [1981], Kyburg [1978a]) persist in ignoring these arguments; thus they attack, not—as they might try to—the idealization as unjustified, but some imagined subjectivism which attempts to avoid any idealization.

(b) *To what Account of Measurement is Subjectivism Committed?*

In claiming idealized betting-behaviour as a suitable measure of subjective probability we need not adopt the naïve operationism which permeates de Finetti's theory. De Finetti [1974] explicitly aligns his enterprise with the general operationist stance of Bridgman: he sees it as a task of redescribing probability-ascriptions in terms of observable behaviour. Since probability is a quantity, this task consists in specifying the procedure for arriving at a numerical value for its magnitude, a procedure which must involve only directly observable behaviour: betting-dispositions.

De Finetti sees as a merit of his approach something which generally presents itself as a problem for operationism about physical quantities—the tendency towards conceptual inflation. If probability is identified with betting-odds, and those odds are just what you, or I, or she would ideally accept, then it makes no sense to speak of the unique probability of an event independent of your, or my, or her probability: which is precisely the subjectivist thesis. No one scale of measurement of probabilities (i.e. set of bets acceptable to an individual) can be said to represent realistically conceived probabilities more accurately than another. Unfortunately, this convenient conceptual inflation need not stop at one probability distribution per individual. Suppose you happen to offer different books on some set of events when you are wearing brown shoes from when you are wearing black. De Finetti cannot avoid accepting as distinct quantities your brown-shoe probability and your black-shoe probability: he cannot make use of a realist notion of relevance to rule out such differences as irrelevant. This fragmentation, aside from its intuitive implausibility, undermines the attempt (which, I shall argue, is central to the acceptability of subjectivism) to make convergence of opinion as evidence accrues compulsory on rational bettors—since each bettor could avoid such convergence by indefinitely subdividing his or her probability quantities.

What is wrong here, it seems to me, is nothing in the subjectivist theory but rather in the unnecessarily strong operationist cast which de Finetti gives to it. I urge instead a modified operationism of the kind which Ellis [1966] advocates for measurement in general. The subjective probabilities of an individual should be identified, not with the bets that individual will accept, but with the ordering relationships of intensity of belief among the individual's partial beliefs. This is not an objective concept, in the sense de Finetti seems to fear, of being independent of the individual. It allows the possibility of more than one method of assessment of subjective probability, but it does not entail that there should be usable methods of assessment other than betting-behaviour. If you do in fact offer different books depending on how you are shod, then your betting-behaviour does not reflect a single ordering relationship. But since, with only minor idealizations, each individual's betting-behaviour is in practice more stable than that, we are entitled generally to ignore such exceptional cases and treat it as a suitable measure, in each person's case, of his or her ordering of beliefs. That in no way drives us away from permitting quite different orderings for different individuals, with no objective external standard of accuracy.

(c) *Problems for Bets as Measures of Degree of Belief*

I have already, in (*a*), drawn attention to three general criticisms of the betting characterization of probability. There are a number of other difficulties which may be thought to vitiate this betting interpretation, of which I intend to examine five, concerning: (i) compulsory bets; (ii) conditional betting; (iii) higher-order probabilities; (iv) bets on generalizations; (v) betting on past events.

One way of dealing with these problems would be to treat them sweepingly in the manner which Blackburn [1980] recommends for dealing with attacks on 'projectivism' in general (including subjectivism about probability). We might see these criticisms as following some such dialectical strategy as: attempt to saddle the subjectivist with a particular view of the meaning of 'probable'; show that view not to correspond with some feature of the meaning we give to 'probable' in certain contexts; and thus conclude that subjectivism is inadequate in that an objectivistic theory is needed for those contexts. My reply would be that the subjectivist does not in fact take the view of the meaning of 'probable' attributed to him or her: (s)he does not repudiate realist talk of probabilities but rather sets out to explain how it is compatible with identifying probability with our subjective evaluations. None the less, it is also quite possible to meet these criticisms without reliance on identifying subjectivism with anti-realism or quasi-realism.

(i) Compulsory bets

For all uncertain events to possess a probability is, in subjectivist terms, to say that we must be prepared to bet on *any* event if challenged to do so. And the subjectivist is just as committed as the objectivist to establishing an interpretation of probability which will extend to all uncertain events. But is it not then implausible to try to measure one's degree of belief in a proposition by the least odds one would accept on its truth if one is compelled to bet? So Kyburg alleges [1978: 164], commenting that 'no odds can be unacceptable to a man who is compelled to bet in any case'.

But this argument goes no distance towards showing the inadequacy of bets to measure belief. All it does is to point up the ineptitude of formulating a measure as the 'least odds acceptable' (which, admittedly, is the formulation adopted in early subjectivist theories). If instead we formulate the measure as the odds an agent could and should choose if betting without knowing the bet's direction or the stake size, the difficulty vanishes.

(ii) Conditional betting

Whatever interpretation of probability one favours, it is clearly essential to establishing a dynamics of belief-change that one assign a clear sense to conditional probability. Several critics have suggested that betting-quotient measures are inadequate because no sense can be made of coherent conditional betting: if I am free to take up your conditional bets when I, but not you, know the truth or falsity of the antecedent, I can so arrange my bets as to ensure a net gain in all circumstances. What conception of conditional bets will rule this out yet not make those bets unsettlable?

This objection can be met by adopting the conception of conditional betting suggested by de Finetti [1972]. We avoid the problem if a bet on q conditional on p is taken to mean a bet which is called off if not p, otherwise won or lost depending on the truth-value of q. This can be achieved by simply laying equal odds on $p \land q$ and against p: then, if not p, there will necessarily be no net loss–gain; if p, the loss–gain will depend on the truth–falsity of q. This notion of conditional bet provides a rationale for the customary characterization of conditional probability as such that: prob (q given p) = prob ($p\&q$)/ prob (p); and by, in effect, defining conditional bets as compound bets of a particular kind rather than bets on a conditional, it escapes at least some of the many problems associated with the truth- or assertibility-conditions of a conditional. (This idea will be developed in §2.1.) For this reason it might, indeed, be preferable to avoid the term 'conditional bet' altogether: rather, conditional partial beliefs are to be seen as measurable by *compound*, but unconditional, bets. This represents, I would claim, at least a terminological improvement—and perhaps, as will be seen later, a conceptual clarification.

(iii) Higher-order probabilities

It may be argued that betting, as a measure of degree of belief, is inadequate because insensitive to higher-order probabilities. I might offer odds of 3 : 1 on a coin landing heads in two quite different situations. First, I might be convinced that it is a normally two-sided coin biased in favour of heads. Alternatively, I might consider it equally likely to be double-headed as two-sided, but in either event fairly tossed (i.e. unbiased in the tossing): I should then offer odds of 3 : 1 on heads too. My betting-odds fail to reveal whether my judgement is purely first-order or involves second-order judgements. (A development of this point yields the 'probability-kinematics' approach—see §3.2. The argument, however, has independent force.)

The drastic response to this difficulty favoured by de Finetti is to repudiate entirely the concept of second-order probability. That seems to me quite misguided: it is surely counterintuitive to rule out allowing my probability-judgements to be themselves the objects of further judgements unless, implausibly, all my personal probabilities are directly open to introspection (so that the only second-order 'probabilities' would be 0 and 1).

A much better answer is available to subjectivism once it has modified de Finetti's strong operationism in the way I suggested earlier, distinguishing an ordering relation from its methods of assessment. The fact that no set of observed betting-quotients guarantees unique identification of an individual's criterion of gradation of partial beliefs is an embarrassment to de Finetti since he defines the partial beliefs as the procedures for assessing them; but, on the weaker view, we may happily take the line that many assessed values are compatible with more than one distribution of degrees of belief, while choosing to presume, until further evidence prevents us, that these values reflect the simplest such distribution. So, in the example quoted, we may suppose, as long as heads keep coming up, that the agent possesses a degree of belief of $\frac{3}{4}$ in heads. If, when tails first comes up, the agent continues to offer odds of 3 : 1 then we can maintain that supposition. But if, as soon as tails first comes up, the agent alters his or her bet instantly to 1 : 1 on heads, we can perceive that we had associated the bets with the wrong ordering relationship (internal to the agent). (For simplicity, I ignore here the conditionalization which would actually occur.) That is no embarrassment to the theory, as it would be if we must identify the betting pattern with the ordering. Indeed, it opens up possibilities of greatly enriching the theory with an account of higher-order probabilities: Chapter 2 will indicate the importance of such a development.

(iv) Bets on generalizations

Some critics argue that subjectivists are willing to bet on the truth of a universal hypothesis, although such bets cannot be settled; this 'reflects the purely hypothetical nature of this betting approach' (Ryder [1981: 166]).

First, not all subjectivists would be willing to bet on the *truth* of hypotheses: certainly not de Finetti, who insists [1974: ii. 201] that subjective probabilities must apply only to 'facts and circumstances which are directly verifiable and of a completely objective, concrete and restricted nature'. Of course, he might, on the basis of a hypothesis, be prepared to bet on the outcome of a trial of the hypothesis.

Secondly, a subjectivist who wants to be free to bet on hypotheses, yet who concedes that bets as commonly understood are readily settlable, need not as a consequence abandon betting-quotients as the central probability concept. Hacking [1965] points out three alternatives: imagining oneself gambling with an omniscient being and unlimited time; denying that generalizations have truth-values, granting them only heuristic usefulness; or using the idea of exchangeability to reduce statistical hypotheses to conjunctions of statements about particular events.

Thirdly, even if there are contexts in which a betting approach could not be applied directly, that does not show it to be useless as an interpretation of probability. One might instead conclude that the term 'probable' should be limited to events to which betting can apply; or that assessing a betting rate was only one method of assessing probability, not always applicable; or that the betting approach often needs to be applied indirectly via exchangeability analyses—and this indirectness is no more serious a problem than the fact that we can only measure the length of a light-wave indirectly.

(v) Betting on past events

Apart from these criticisms of Ryder's, another argument against, or at any rate in favour of limiting, subjectivism is advanced by L. J. Cohen [1977] when considering the judicial context. Cohen complains in general about 'the absurdity of talk about accepting reasonable bets on unsettlable issues' [1977: 91]; such issues comprise both unbounded generalizations, which we have just considered, and also judgements about the past, including the sort of judgements juries are asked to make. The problem is this:

Knowing nothing, or only a little, about the local archaeological evidence I can wager you, on the basis of evidence elsewhere, that there was no Roman settlement at Banbury; and to settle the bet we can excavate and see. But, if all the appropriate excavations have been done already, and we know their results, there is nothing to wager about. Similarly, to request a juryman to envisage a wager on a past event, when, ex hypothesi, he normally already knows all the relevant evidence that is likely to be readily obtainable, is to employ the concept of a wager in a context to which it is hardly appropriate. [1977: 90]

And yet, in these judicial and historical contexts we do make probability-judgements, although, if this argument is correct, no betting could occur.

There is no very easy way out of this difficulty. A moderate personalist could agree that some interpretation of probability was appropriate
in these contexts other than subjective degree of belief: but that option
is closed to the strong subjectivist. Even if an argument could be sustained that there is a gap between verdicts and beliefs in the judicial
context so that verdicts need not embody probability-judgements, that
will not help in the case of judgements made by the archaeologist or
historian. We might suggest that in such cases it is not certain that bets
are not settlable: if you bet me at odds of 7 : 3 today that there was a
Roman settlement at Banbury and tomorrow God reveals to us both that
there was not, I should expect to collect on the bet. But this is a feeble
retort. It remains unreasonable to bet on events where a settling event,
if logically possible, is yet enormously unlikely—at the very least it is
unreasonable to found an account of belief in such events on such an
etiolated notion of settlement of bets.

It seems that, in order to meet this objection, the subjectivist must—
as I have tried to avoid in (i)–(iv)—align his or her position with, or at
least develop it along similar lines to, some form of anti-realism about
the past. Statements ascribing a probability-value to the existence of a
Roman settlement in Banbury must then be interpreted as judgements
of the degree of belief presently warranted by evidence presently available to us: these judgements are measurable by bets in so far as two
agents can establish between them whether a judgement one has expressed is in line with all the items of evidence and the links between
them and the agent's other beliefs. Major questions then arise—which
lie beyond the scope of the present work—as to the link between the
probability-value of the past-tense statement 'With probability p there
was a Roman settlement here', made presently, and the truth-value of
the present-tense statement 'There is a Roman settlement here', made
at the appropriate past time.

1.3. PROBABILITY: DEFINITION AND FUNDAMENTAL THEOREMS

The general considerations of the previous sections now make it possible to formulate more precisely the concept of probability. Mathematically, this is worked out in great detail by de Finetti in the two
volumes of *Theory of Probability* [1974]: here I offer only a bare summary, but one that should be adequate for the philosophical discussion

of the rest of this chapter. (Howson and Urbach [1989: chs. 2 and 3] provide a mathematically and philosophically lucid account in moderate detail.)

The opinion of each individual faced with a situation of uncertainty is to be represented by a *prevision-function P*, so that for each random magnitude X there is a value P(X), *the prevision of* X. In the case of events the prevision-function P becomes a *probability distribution* over those events, and for any event E the prevision of E becomes the *probability of* E, p(E), for the given individual. To arrive at an operational definition of prevision/probability we must seek a means of reducing these concepts to observable dispositions in an idealized betting situation.

(In the previous section I presented bets indifferently in terms of loss, gain, or price of goods assumed to be of positive utility. Formally, it makes no difference whether one defines $P(X)$ in terms of losses or gain, penalties or prizes. In deference to the historical origins of probability theory in gambling, where the bank ultimately wins, de Finetti chooses to use the terminology of loss for his principal definition, while also providing a somewhat intuitively clearer equivalent definition in terms of gain.)

The loss definition of $P(X)$ is arrived at thus: suppose you are an ideally rational agent—de Finetti dignifies this creature as 'You'—and You are to suffer a penalty L, equal to $\{(X - \bar{x})/k\}^2$ that is proportional (k = constant) to the square of the difference between X and a value $\bar{x}$ which You are free to choose. Then $P(X)$ is the value of $\bar{x}$ which You would choose for this purpose. Equivalently, in terms of gain, suppose You are obliged to estimate X at a value $\bar{x}$ on the understanding that, after making this choice, You are committed to accepting any bet whatever (in either direction) with gain $c(X - \bar{x})$, where c is an arbitrary constant which your opponent may determine. Then $P(X)$ is again defined as the value $\bar{x}$ You would choose for this purpose. When X is an event E, inf $E = 0$ and sup $E = 1$, so that $p(E)$ is then a normed measure defined over the interval [0, 1] (the endpoints of which are the possible values E can take, non-occurrence and occurrence), such that $p(E)$ is the certain gain which You judge equivalent to a unit gain conditional on the occurrence of E.

What we have so far is of course not yet adequate as a definition of probability, if we wish these measures to conform to the axioms of the probability calculus. You have been allowed to choose any $\bar{x}$ You wish, so all this value so far indicates is the betting-odds You are offering. In particular, You may be betting in such a way that, given a set of events E_i, $(i = 1, 2, \ldots n)$, which are mutually incompatible, Your values of $\bar{x}_i$ may be such that for You $\bar{x}_E \neq \sum \bar{x}_i$ although E is the union of $E_1, E_2 \ldots E_n$. So, although Your values of $\bar{x}$ satisfy two of the axioms of the probability calculus, they fail the third axiom of additivity (even if it is interpreted as requiring only finite additivity). Hence this measure $\bar{x}$ is not as it stands a probability-measure.

To make it into a probability-measure, it is necessary to impose a condition on Your choice of $\bar{x}$—the condition of *coherence*. Any set of Your previsions is said to be coherent iff among the combinations of bets to which You are committed there are none for which the gains are all uniformly negative; there is no subset which Your opponent could choose which would with certainty result in a loss for You, giving rise to what is often called a Dutch book. The remarkable aspect of this coherence requirement, which Ramsey and de Finetti were the first to make explicit, is that

it constitutes the sole principle from which one can deduce the whole calculus of probability; the calculus then appears as a set of rules to which the subjective evaluations of probability by the same individual of various events ought to conform if there is not to be a fundamental contradiction among them' (de Finetti [1974: i. 84]).

For it follows very simply from the definition of coherence that it is a necessary and sufficient condition of Your conforming to the probability calculus that those beliefs be coherent (we may call this the *Dutch book theorem*); further, if You are confronted by a complete class of incompatible events, it is a necessary and sufficient condition of Your probability-judgements conforming to the calculus that the sum of these probabilities be 1 (de Finetti calls this the *theorem of total probability*).

In his major work [1974: 112], de Finetti formalizes these results into his *fundamental theorem of probability* which, in summary, states that:

given the probabilities $p(E_i)$, $(i = 1, 2, \ldots n)$, of a finite number of events, the probability $p(E)$ of any further event either

(a) is uniquely determined if E is linearly dependent on the E_is or
(b) can be assigned coherently any value in a closed interval $[\acute{p}, \acute{p}']$ where $\acute{p}$ is the supremum of the evaluations from below of the random quantities X linearly dependent on the E_is for which we certainly have $X \leq E$ and $\acute{p}'$ is the corresponding infimum.

This theorem permits us, even where the number of choices is infinite, to proceed to attribute to all the events we wish, one after the other, any probability-values coherent with the preceding ones.

Thus far, all the probabilities of which we have been speaking are relative to an individual's state of information. But it is still necessary to develop a concept of the probability of an event E conditional on an event H, in order to be able to combine probabilities which are relative to different states of information, as is essential if we are to consider statistical or inductive inference. It can be proved (de Finetti [1974: i. 136–40]) that a necessary and sufficient condition of one's conditional probabilities being coherent is that they satisfy

the *theorem of compound probabilities*, $P(H \wedge E) = p(H) \times p(E|H)$, or its corollary,

Bayes's Theorem, $p(E|H) = \{p(E) \times p(H|E)\}/p(H)$.

As Chapter 2 will show, these latter theorems are fundamental to a subjectivist account of intersubjective agreement.

1.4. COHERENCE, CONSISTENCY, AND DUTCH BOOKS

(a) *Consistency and Coherence*

As has been seen, de Finetti [1974: i. 84] identified the central feature of the subjective interpretation of probability as its imposition of only one, minimal, constraint upon probability-judgements—the requirement of coherence, to which 'subjective evaluations of probability by the same individual of various events ought to conform if there is not to be a fundamental contradiction between them'. In this section and the next I shall explore the force and scope of this constraint and its connection to subjectivist conceptions of rationality in probability-judgement.

Whatever the later variations in formulation of the concept, it is clear that coherence of probability assessments is to be treated as analogous to, or even as a generalization of, deductive consistency. This analogy suggests some preliminary, but important, considerations.

First, it might seem plausible to identify full demonstrability with the limiting cases of assigning probability 0 or 1 to a proposition. Subjectivist theories can accommodate acceptance of a close resemblance between probability and provability, with coherence providing a constraint upon partial beliefs parallel to the constraint imposed upon full beliefs by consistency. But it would be rash to move from analogy to assimilation. Ellis [1973: 129] argues that 'our logics of truth and of certainty should coincide': if we adopt as our logic of certainty a probability system A with the range of values restricted to 0 and 1, and as our logic of truth a standard first-order propositional calculus B, A and B should 'correspond' in that for all theorems α, β, γ ... of B, $p(\alpha) = 1$ will be a theorem of A iff α is a theorem of B, and $p(\alpha) = 1$, $p(\beta) = 1$, ... $\vdash p(\gamma) = 1$ will be valid in A iff α, β, ... $\vdash \gamma$ is valid in B. Conditional probabilities, Ellis claims, do not satisfy this correspondence principle: hence we need either a weaker logic of certainty or a stronger logic of truth. I shall argue later that Ellis's account of conditional probability is mistaken. But, in any case, his initial assumption that our logics of truth and certainty should correspond is problematical. Suppose I am throwing darts whose tips are geometrical points, at random, at a dartboard. Then I would assess the probability that the dart will strike a specified geometrical point on the board as 0—that is, I should accept any odds offered, however unlimitedly unfavourable, against its striking that point. But I am not thereby committed to claiming the truth of the proposition that it will not strike that point. Preparedness to stand an unlimited loss if an infinitely unlikely, but 'non-ignorable', event should occur need not entail preparedness to accept that one has made a mistaken claim to knowledge. It is an important merit of subjectivism that, in thus distinguishing prevision from prediction, it allows—especially, though not only, in cases of zero probability which objectivist theories find paradoxical—for distinguishing two types of commitment: to risking loss as against risking a falsified claim to knowledge. Rash identification of probability 1 with truth-claims would discard this advantage.

Secondly, it is clear that de Finetti, presenting coherence as a generalization of consistency, intends thereby to prepare the ground for his version of strong subjectivism in which coherence is the *only* constraint on rational probability-judgement: like consistency, it is the paramount regulative ideal, sufficient alone to guarantee rationality if its demands be met. But this rock upon which de Finetti bases his system is perhaps not as secure as he imagines. Rationality may demand that one accept that at least one of one's beliefs in some connected network of beliefs is mistaken—this is the nub of the preface paradox: where the beliefs are logically related to one another, we must accept as a sound generalization that, in addition to some beliefs being wrong, there will be some inconsistencies among them (though of course we do not know

where these occur). Rationality appears to endorse treating consistency
as the principal merit in our theories without insisting that it always
dominates all other merits: we should not be content to achieve consistency by reducing our beliefs to one atomic proposition, for example,
any more than to achieve coherence by setting all our betting-quotients
at 0—in effect, refusing to bet—except for an isolated wager. The
arguments against treating coherence as a sufficient constraint for rationality in probability-judgement cannot be evaded by appealing to
analogies with consistency.

These points represent, not so much criticisms of the concept of
coherence, as caveats about attempting to rely on analogies with consistency to supply what should instead be part of the constructive theory.
How the theory meets these demands will become apparent in the next
section and in Chapter 2.

(b) *Coherence, Dutch Books, and the Probability Calculus*

Why is it undesirable for our partial beliefs to be incoherent? Ramsey
in 1926 was the first to see that incoherence will leave the bettor open
to being forced by a shrewd opponent into inevitably disadvantageous
bets.

Later subjectivists, from de Finetti on, have turned their attention to
using this situation of certain loss as a basis for defining coherence of
probability-judgements. We have seen that de Finetti defines a set of
bets as coherent if 'among the combinations of bets which You have
committed Yourself to accepting there are none for which the gains are
all uniformly negative (giving rise to what is called a "Dutch Book")'
[1974: i. 87]—that is, no betting arrangements where you are bound to
lose no matter what the actual outcomes of the events. Other subjectivists
have extended the idea of coherence to encompass, in addition to de
Finettian weak coherence, a definition of strong or strict coherence as
immunity to semi-Dutch books—combinations of bets where you may
lose but cannot win.

The significance of de Finetti's early work (especially [1937]) lies
in its proofs that from the single requirement of coherence, defined as
avoidance of a Dutch book so presumed to be obviously desirable, can
be derived the requirement that one's beliefs correspond to the probability-measures of the formal calculus. It is a necessary and sufficient
condition for one's betting-quotients to be coherent that they should be
treatable as probabilities satisfying the axioms of §1.2—they must be

additive (although, unlike in Kolmogorov's full axiomatization, they may be only finitely additive), complementational measures mapped onto [0, 1]; further, if one is betting on a complete class of incompatible events, it is a necessary and sufficient condition for coherence that the sum of one's betting-quotients be 1 (the theorem of total probabilities).

It is worth noting, since it seems to be the source of some confusion (see e.g. Mellor [1971: 41]) that de Finetti defines coherence rather differently from most personalists (such as Jeffrey) and personalist critics of subjectivism (such as Kyburg). The basis of any subjectivist theory must be that avoidance of Dutch books entails, and is entailed by, conformity to the formal calculus. Standardly, coherence is taken to mean conformity to the calculus, so that de Finetti's fundamental theorems are taken as establishing that iff we are coherent we avoid Dutch books. But, as we have seen, de Finetti himself attaches a different meaning to coherence, *defining* it as the impossibility of a Dutch book, so that his theorems mean that iff we are coherent we must be in conformity to the calculus. This is not merely a terminological point but a genuine difference of emphasis. For, coherence is taken also to be a criterion of rationality analogous to consistency. So, while the standard development of subjectivism runs from consistency through conformity to the calculus to rational action, the theory I am advocating runs from consistency through rational action to a justification of the applicability of the calculus. The importance of this variation in what might be termed the methodological place of coherence will appear in considering the normative force of coherence (§1.5) and in comparing the plausibility of competing personalist theories as accounts of rational judgement (Ch. 3).

Thus far I have dealt only with the coherence of unconditional probability-judgements. But it is vital to a strong subjectivist programme that the concept of coherence can be extended to cover conditional probabilities. Since, within the approach I am urging, coherence is identified with immunity from Dutch books, it is essential to show that Dutch book arguments apply to conditional probabilities—which, in turn, means that conditional bets must be brought within the scope of the coherence requirement. (The necessity to define dyadic coherence—though not the definition I will adopt—seems to have first been recognized in the early 1970s by Lewis: see Teller [1973].)

This burden on apologists for subjectivism has been used to provide arguments against the subjectivist thesis by, among others, Ellis, Mellor, and Kyburg. Their arguments share a common structure: for subjectivism to make sense of some important aspect of probability-judgement it needs to be able to define coherence of conditional probabilities; it cannot do so; hence it is incomplete or inadequate as a probability interpretation. These arguments fail, I believe, because although the first premiss is true and the argument-form valid, the second is false.

There are three main arguments advanced in support of the first premiss, of which the last two seem to me to be successful.

First, Ellis [1973] argues that we need coherence of conditional probabilities in order to make sense of the correspondence of our logics of truth and certainty. But, as I pointed out earlier, such correspondence is largely illusory; certainly, it is not a necessary feature of subjectivism.

Secondly, both Ellis [1973] and Mellor [1971] argue that most scientific generalizations and undecided historical hypotheses have in common with the objects of conditional bets that they are semi-decidable—that is, they could become accepted as true, could become accepted as false, or could just remain undecided. If coherence cannot be defined for conditional bets then subjectivism cannot provide a criterion regulating judgements of the probability of such hypotheses and generalizations.

Thirdly, both Mellor and Kyburg [1978 and elsewhere] argue that, unless coherence can be extended to conditional probabilities, it will remain essentially a static constraint, unable to account for convergence of judgements as evidence accrues. Moreover, such a limited constraint would force us to accept as rational any change of an agent's judgements from one coherent state to another regardless of the process which may have been adopted to change them—such as, Kyburg suggests, consulting one's parrot. But we want from any complete account of probability an explanation of what makes some changes in our systems of partial beliefs, and not others, rational.

There are two main arguments for the second premiss above, neither of which seems to me to be successful.

First, Mellor posits a gambler whose partial beliefs in some proposition q conditional upon two mutually compatible propositions r and s are unequal. Now suppose both r and s turn out to be true. The gambler is committed to betting on q at two different quotients, since $p(q|r) \neq p(q|s)$ and, if conditional betting is to make sense at all, one must remain committed to the bet after discovering the truth of the antecedent. But in betting at these different quotients the gambler will be liable to a Dutch book. Hence Dutch book arguments do not apply to conditional betting-quotients.

But this argument is vitiated by positing an initial situation which, if conditional coherence can be defined, will thereby be ruled out. Grant that Dutch books do apply to conditional betting-quotients. An initial, though simplistic, response would be to claim that all Mellor's arguments show is that anyone who assigns different values to $p(q|r)$ and $p(q|s)$ where r is compatible with s is incoherent, and that to advance the mere possibility of a gambler's accepting such different odds as demonstrating the impossibility of defining coherence is to beg the question. This is obviously inadequate as it stands: suppose, for instance, that q is 'A develops lung cancer', r is 'A is a heavy smoker', and s is 'A has a family history of cancer'—a coherence requirement that $p(q|r)$ must equal $p(q|s)$ seems ridiculous. Nor is it much help to evade the problem by insisting, as de Finetti [1974] did, that the proposition on which the bet is to be conditional must state the *only* relevant fact which could become known: that is thoroughly counterintuitive as a description of everyday judgements. The solution is, rather, to identify $p(q|r)$ either with the strength of an ideally rational agent's partial belief in q given only r as evidence or, more realistically, with the strength of that agent's partial belief in q, relative to a corpus of beliefs including belief that q, given that only r is to be added to that corpus. I develop this idea further when considering weight of evidence in §2.4.

Secondly, Ellis distinguishes Dutch books from what may be termed (following Shimony) semi-Dutch books; since conditional bets may remain undecided they can be governed only by the latter; if they are governed by the latter constraint then the quotients offered in the bets cannot be probability-measures. So, Ellis concludes, conditional probabilities cannot be governed by a coherence requirement derived from Dutch book arguments. But this conclusion rests upon a false premiss. What the argument shows is that if conditional bets are to be subject to a strict coherence requirement then they will not correspond to quotients treatable as probability-measures. But it remains open that conditional bets should be governed by the weaker de Finettian coherence constraint: that is, the coherent bettor cannot possess a system of conditional beliefs such that a subset can be chosen for betting which will lead to certain loss. Ellis himself admits this possibility in conceding [1973: 134] that 'it may be possible to avoid this conclusion by . . . adopting the non-standard rule that if any constituent bet in a book remains undecided the whole book remains undecided and all stakes are returned'. For, in place of that non-standard rule in a system subject to strict coherence, one may achieve the same result in a system subject only to weak coherence by adopting the rule for conditional bets (treat them as compound bets) which I outlined earlier.

This idea will become of especial importance when, in Chapter 2, we come to consider belief change via Bayesian conditionalization.

(c) *Criticisms of Dutch Book Arguments*

(i)

It has been suggested that no constraint based upon avoidance of Dutch books can serve as a criterion of rationality for subjective probability, because a non-coherent betting policy might be quite rational for a non-acquisitive man. Two criticisms of Dutch book arguments follow.

First, it would seem that even a totally non-acquisitive individual, who attached no value to anything, could hold partial beliefs which should be interpretable as probabilities: but how can betting have any meaning for such a man? Once again, a perfectly adequate subjectivist response (which de Finetti's naïve operationism closed off to him) is that subjective probabilities represent an ordering relationship which is in most cases measurable by betting-quotients, but that it does not damage the theory that it may, or even that it should, be possible to attribute partial beliefs in extreme cases where betting is inappropriate as a measure—our anomie-ridden man, or perhaps some animals.

Secondly, suppose the non-acquisitive man to be someone possessed of values which, however, are not represented by price, loss, or gain as for most of us: if, say, his values encompass deriving spiritual benefit from relinquishing his possessions, might he not rationally do so by allowing more materialistic gamblers to make Dutch books against him? (Cf. the rich young man in the Gospel story.) Certainly he might. But all that shows is that coherence must be taken to apply to betting as a communal activity (for a community of at least two people): if there is such a thing as private betting, coherence makes no sense for it. Further, the parties to the bet must be presumed to have an agreed scale of values, though we need not be able to specify what that is to consist in. Two such spiritually minded individuals might bet, as they might play a game of losing chess with the intention of winning it; but one of them could not be said to be betting merely because he was exchanging goods with someone who did not share his value-scale. The argument shows that any subjective theory which idealizes from actual to coherent behaviour must presume a shared set of utiles. It does not show that such an idealization must be faulty.

(ii)

The concession which subjectivism has had to make to this argument, however, does leave it open to a more damaging criticism. As I argued in §1.1, de Finettian subjectivism avoids several of the problems beset-ting preference-based theories by taking utility as primitive: it develops a unique criterion of gradation for probabilities assuming, not that there is a unique scale of utility, but that there is in any gamble *some* agreed scale of utility relative to which loss and gain may be defined. So far, the arguments of the previous paragraph cause no difficulties.

Where they do bite, however, is into the suitability of coherence as a criterion of rationality for betting behaviour. Why should we be co-herent?—because it is obviously undesirable to be forced into losses: as Kyburg and Smokler [1964] put it, it is tautologous that we do not wish to lose objects of positive utility under all possible circumstances. But suppose, as is being suggested, that we know no more when betting than that there is some shared scale of value between us and our op-ponent. How will we then recognize the outcome of the bets as undesir-able? Surely we need to be already equipped with a criterion determining which outcomes possess positive, which negative, utility in order that the concept of a Dutch book should hold any meaning for us. Then, if Dutch books presuppose that utilities are already in place, it becomes impossible to run the theory of probability in the direction de Finetti intends without simultaneously proffering an account of utility. But now we are back with the problems surrounding preference theories.

Subjectivist theory has as yet no more than the outlines of an answer to this objection. Certainly, de Finetti nowhere provides a solution. I suspect that the answer is to be found in relinquishing the, very conven-ient, assumption that one can develop a utility-independent concept of probability for all judgements directly from Dutch book considerations. Instead, we may need to put more weight on Ramsey's idea that a probability interpretation in terms of partial beliefs should initially be established only for a basic fragment of our judgements—those towards which we are indifferent as to their truth or falsity. As I indicated earlier, we need only be able to recognize two states of affairs α and β for which we prefer one to the other, to enable us to generate an order-ing of our degrees of belief within the basic fragment. Once we have achieved this much we need then acquire only one further preference-relation—preference between 'α if S, β if not S' and 'α if P, β if not P' for each S outside the fragment and some P within it—to enable us to generalize the theory to all our probability-judgements.

This answer concedes that a limited notion of utility is necessary in order to define coherence but tries to minimize its complexity. Clearly, though, the de Finettian approach would be more comfortable if it could be defended without resort to such a damage limitation exercise.

(iii)

The objection has sometimes been raised (e.g. in Ryder [1981]) that coherence, in guaranteeing an individual against suffering a Dutch book, offers no protection against finding oneself part of a group against whom a Dutch book can be made. It is easy to see that any bettor who can find two or more opponents with differing, albeit coherent, distributions can arrange his or her bets so that the group will always lose to him or her. But this does not invalidate the use of Dutch books to define rational action for an individual; it merely shows that, if we wish additionally to define group coherence (which generally is not needed) we must define it in terms of communally agreed betting-rates. The argument can only appear to provide a criticism of Dutch book arguments by fallaciously sliding from treating the members of the group as individuals when laying the bets into treating them as one entity when counting the winnings.

(iv)

When this objection is combined with the requirement that not only one's unconditional but one's conditional bets be coherent it does produce a substantial problem for a de Finettian theory. Granted that none of us are perfect logicians, there are bound to be gaps or divergences in our conditional judgements which will enable an opponent, who knows no more than we do about events in the world but is a better logician, to make a Dutch book against us in just the same way as against a group which had failed to agree on a communal betting-quotient. Suppose, in the summer of 1993, I had been speculating on the selectors' likely choice of players for the forthcoming Test series in the Caribbean. Quite plausibly, I could in practice ensure that my unconditional betting-quotients were coherent; even in more complex situations, it requires no very great idealization to suppose me capable of unconditional coherence. But what of the enormously large network of conditional bets which I must be prepared to stand: assessing the probability of, say, Gatting being selected if Gower tours and if Gower does not tour, the probability of Gower touring given that the selectors are sober, etc.? The definition of coherence allows that a coherent bettor may lose to someone who is factually better informed but not to someone who is merely better at perceiving all the possible logical links among the bettor's conditional probabilities.

There are at least three possible lines of reply to this objection.

First, one might bite the bullet of idealization and accept that the coherence requirement governs behaviour which is not only ideally rational but logically omnipercipient too. The cost of such acceptance would be greatly to strain any claim that norms of coherence can be extracted from actual behaviour while retaining more than a very tenuous link to it.

Secondly, one might take the radical step of distinguishing partial beliefs from assent to those beliefs, or, perhaps, assent from manifestation of assent. That might well solve the problem of excessive idealization—but at too high a cost. It would then become much too easy to be coherent, simply by being very guarded in one's utterances. Coherence works as a constraint because it is a constraint on beliefs which we may at any moment be required to display—not on voluntary patterns of assent.

Thirdly, one might attempt to prevent the combinatorial explosion of conditional bets by setting limits to what may rationally be required of a bettor as to which propositions (s)he will accept as evidence relative to which others may be conditionalized. Both Lehrer [1972] and Shafer [1978] have argued that the subjectivist de-emphasis of the notion of evidence will create problems not only for the definition of coherence (though they have in mind only unconditional coherence) but also for accounting for change of partial beliefs based on a change of view as to what to count as evidence for some proposition. Subjectivism needs to, and can, encompass an account of how sentences compete for evidential status.

(v)

Finally, let us turn to the criticisms of the idea of coherence which stem from assessing the reliability, or degree of calibration, of sequences of probability-judgements. Suppose you are a weather forecaster offering a long sequence of judgements, day by day, about the probability of rain the next day. Assume that you are coherent. Assume that your sequential judgements are made with feedback—that is, each judgement is a dyadic judgement given the precipitation outcome of all the previous days. Assume that your total coherent distribution Π includes a judgement about your own calibration performance, in the following sense: select all those days for which your forecast probability p was, say, 0.01; determine the (long-run) relative frequency q of those days on which it in fact rained; repeat for $p = 0.02, 0.03 \ldots$ and plot p against q. If the graph is the diagonal $p = q$, you are perfectly calibrated. Degree of calibration could be measured by any variance measure such as sum-of-squares-of-deviations or a correlation coefficient.

It is possible to prove an extremely interesting theorem about calibration. Assuming only that your forecasts are made sequentially according to a fixed probability-distribution Π it can be shown that, with Π-probability 1, as the sequence of forecasts tends to infinity then $q - p$ will tend to zero. This theorem does not, of course, imply that *any* coherent forecasts will be well calibrated. The key clause in the theorem is 'with Π-probability 1', which refers the result to the only distribution being considered, that of the forecaster. In other words, if you are coherent you must assign probability 1 to the event that you are well calibrated, even though you in fact may not be. The worrying corollary is that you must then assign probability 0 to the event that you are malcalibrated; so, the apparently very weak constraint of coherence seems to compel you to an unrealistically strong faith in your own calibration level. Most of the charges commonly brought against strong subjectivism attack coherence as too permissive a constraint on its own for it to be possible to base an interpretation of probability on it. The calibration paradox, however, threatens instead to show that coherence will force us into unreasonably strong commitments.

The result appears to have been discovered first in 1962 by Pratt, a Harvard statistician, who never published it. It was rediscovered [1982] by the London-based statistician Phillip Dawid, who proved it in very much the form I have just described. Since then, Seidenfeld [1985] has published a more general proof which extends the result to infinite feedback spaces and to cases where the forecaster's personal probability is not countably additive: Dawid's and Pratt's theorems come out of this general proof as special cases. Most cases of practical interest will not require these generalizations, so let me continue to focus on Dawid's version of the theorem, which I now state a little more rigorously.

You contemplate making a sequence of forecasts, Π, about the events e_i, rain on day i, $(i = 1, 2, \ldots)$. You receive feedback information f_i subsequent to the ith forecast and prior to the $i + 1$st. This feedback f_i is sufficient to determine the outcome of your ith prediction, e_i; so we must always be able to extract from f_i a binary indicator

$$I_i \ (= 1, \text{ if } e_i$$
$$(= 0, \text{ otherwise.}$$

In general, then, the probability you announce for day i is $p_i = p \ (e_i | f_1, \ldots f_{i-1})$.

Consider an arbitrary, infinite subsequence identified by the sequence $\delta_i \ (i = 1, \ldots)$ where δ_i is either 0 or 1. Now define:

$$v_k = \Sigma \delta_i \ldots \tag{1}$$
$$rf_k = v_k^{-1} \Sigma \delta_i \ I_i \ldots \tag{2}$$
$$p_k = v_k^{-1} \Sigma \delta_i \ p_i \ldots \tag{3}$$

all summations being over $i = 1$ to $i = k$. In words, (1) is the number of forecasts chosen from among the first k to form the test subsequence; (2) is the relative frequency with which events e_i occur relativized to those tested by the subsequence among the first k forecasts; (3) is the average forecast probability within the tested subsequence over the first k forecasts.

Dawid's theorem now states simply:

With Π-probability 1, as $k \to \infty$, $rf_k - \bar{p}_k \to 0$.

The proof of the theorem is a straightforward application of a standard theorem in probability, the law of large numbers for martingales. (A martingale is a probability-function on an arbitrarily long sequence of events such that all bets on any subsets of the sequence are fair, i.e. have expectation zero.) The results quoted at the end of §1.3 (together with the analysis of conditionalization to come) commit the subjectivist to that law as a direct consequence of coherence and a diachronic conception of conditionalization. So, from the minimal rationality constraint of coherence, we derive the requirement that we must attach probability 1 to our being perfectly calibrated, hence probability 0 to our being at all malcalibrated. Yet it seems entirely irrational to possess such total confidence in our own forecasting performance, particularly since we will never have such confidence in anyone else's.

There is no simple solution to this problem, although several relatively brisk responses might be contemplated.

1. One rapid way to dismiss the problem would be to rule out higher-order personal probabilities entirely. This would need to be done on philosophical grounds rather than mathematical: there is no mathematical difficulty in treating probability as itself a random variable. As I pointed out earlier, de Finetti seems to take this line.

He writes [1972: 32]:

Any assertion concerning probabilities of events is merely the expression of somebody's opinion and not itself an event. There is no meaning, therefore, in asking whether such an assertion is . . . more or less probable.

And later [1974: i. 114], he claims that talk of probabilities of probabilities resembles, when conducting a survey, recording not only the 'don't knows' but those who don't know if they are 'don't knows', which is patently absurd.

The analogy is not convincing as it stands, and in any case the survey method is not that patently absurd. Someone (it may have been Arne Naess) once conducted a survey among philosophers to assess how much agreement existed in categorizing a list of sentences into analytic and synthetic—hard cases mostly, like 'summer follows spring'. Respondents could reply that a sentence was analytic, or synthetic, or they didn't know. It would be very interesting, in the light of conflicting positions about the analytic–synthetic distinction, to know how many would claim that they knew they didn't know.

Nor is de Finetti's dismissal of higher-order probabilities as meaningless tenable. Suppose that one Monday in three, at random, I become so bored with assessing probabilities that I take the outcomes of all binary guesses as equally probable. Cannot I now meaningfully, even justifiably, assert that the probability that my personal probability next Monday, 7 December, that the following day's noon temperature will be over 100°F is $\frac{1}{2}$, is $\frac{1}{3}$?

2. Perhaps there are coherent higher-order probabilities but they are always trivial: they take on the values zero or one only (so the example I've just given is a meaningful judgement, but could never form part of a coherent set of judgements). A bad argument for such triviality is that partial beliefs are introspectively transparent—you always can be certain whether you have them or not. If partial beliefs are dispositions to act in a particular way—say, to bet at some odds rather than others—it is surely entirely possible that you do not know your own mind with certainty. A much better argument emerges out of the nature of coherence. A coherent set of bets is a distribution of risks which cannot encounter a Dutch book. If you are challenged to bet that your bets are wrong, you must put the odds on this at zero. If you did not, you could always avoid a Dutch book by laying off your bets against yourself: so, if coherence is to constrain judgements at all, bets on the rightness of bets must be avoided—that is, probabilities of probabilities must vanish

into the limiting-values of 0 and 1. A slightly different version of this argument appears in van Fraassen's [1984] paper, 'Belief and the Will'. If you assign any value other than 0 to the probability that your future probability of A will differ from your current dyadic probability of A given that future value, you will be incoherent—a Dutch strategy can be set up against you. Either way, calibration comes out as a special instance of this triviality result for second-order probabilities.

There is something of value—as well as much that is mistaken—in these arguments, and I shall return to them later. But one thing they will not achieve is a dissolution of the calibration problem. Rather, they extend it. For it now seems that we must, to be coherent, express total confidence in any coherent set of our judgements, sequential or not, with feedback or not, with or without information about relative frequencies. This cure is worse than the disease.

3. We might note that the calibration problem is structurally similar to the preface paradox for full beliefs. Could we attempt a similar resolution, so that, for either partial or full belief, belief in the conjunction of a set of propositions is not entailed by possessing the corresponding conjunction of beliefs in each member of the set?

Perhaps. But then we will no longer be able to treat partial beliefs as measurable by betting-quotients; for the outcome of a conjunction of bets must be the conjunction of the outcomes of each bet. Then it becomes very difficult to see how partial beliefs can serve to interpret probabilities—quantification of degree of belief is rendered impossible.

4. Kadane [1982] has suggested that we need not worry about the calibration theorem since it concerns only what will happen in the infinitely far future: it has no bearing on any judgements about finite sequences.

This is totally mistaken. First, the theorem does not say, as Kadane claims, that 'in the infinitely far future I believe I will learn every-thing . . . about whether it will rain tomorrow', if only because the syntax of such a claim would be very shaky. It says instead that I must believe, now, that in the infinitely far future I will learn everything about whether it will rain tomorrow. So its consequences are immediate for my current judgements, albeit current judgements about the long term. Secondly, the theorem (like the martingale convergence theorem) is a strong, not a weak, law of large numbers. That is, it asserts not only that for every large n, the value of rf_n is likely to be near $\bar{p}_n$, but also that rf_n converges onto $\bar{p}_n$: given any $\varepsilon > 0$, we can choose a value of n large enough so that, with probability 1, $rf_n - \bar{p}_n < \varepsilon$. In other words, what we must believe now is a convergence result entailing constraints on our probability beliefs in the far, but not infinitely far, future.

At the opposite extreme to these brisk efforts to dismiss the problem, the calibration theorem seems to have had an intoxicating effect, producing some overexcited assessments of its significance.

5. An initial temptation for a staunch subjectivist would be to suppose that the theorem shows that, if you are coherent, you actually must be well calibrated, and that we now possess a quick and simple proof that objective probability is redundant. Of course, the theorem deals with personal probabilities within an agent's distribution, so that the clause in the theorem 'with probability 1' regulates only internally, constraining the degree of belief the agent can have.

6. Crude as this error is, traces of it seem to linger on in van Fraassen's paper [1983] (as the title 'Calibration: A Frequency Justification for Personal Probability' indicates) and, perhaps, in Seidenfeld [1985]. Van Fraassen hopes to use calibration as a frequency-theoretic justification for imposing the constraints of the formal calculus on one's partial beliefs—a frequency-theoretic justification which will replace the Dutch book justification. He takes calibration to be a form of agreement between a probability distribution and long-run frequency, agreement which can be shown to be possible only if the distribution conforms to the formal calculus. The mistake here lies in confusing external correspondence between a probability distribution and frequencies, with the internal constraints determining the value which may, or must, be assigned within that distribution to the proposition that such a correspondence obtains. Seidenfeld seems to half-advance a similar position [1985: 283]: 'If long-range calibration is thought to signal realism, as an objective validation of subjective beliefs, then mere coherence is sufficient to establish (almost certain) agreement with the facts, in the long run . . .'—which he then retracts: 'The agreement is "internal" . . . as the result holds . . . from the point of view of the forecaster.' The retraction is, as we have seen, compelled by the second-order nature of the theorem. If we feel in need of a frequentist justification of subjectivism, it must lie in the first-order convergence results established by de Finetti and Savage, not in the notion of calibration.

7. Pretty much the diametrically opposite view is taken of the calibration theorem, in the [1982] Dawid paper—that it has 'destructive implications for coherence' and, it is implied, for the entire subjectivist programme. As I have argued earlier, that pessimism is overstated.

There is a problem here for any probability interpretation, even if it is worse for subjectivism than for its competitors. But, at worst, failure to resolve it leaves us with a foundational anomaly, not a reason to abandon an otherwise successful theory for some competing theory beset by the same anomaly.

8. There is a suggestion made initially by Lindley [1982] which seems to offer a solution to the problem based on the very tolerant definition of coherence which we have been using. Coherence requires that you do not regard as fair a bet whose return is certain to be negative whatever the outcomes of the events involved. Kemeny [1955] and others have criticized this definition as too weak, urging instead that we adopt a principle of strict coherence which refuses to allow as fair a bet whose return is never positive and sometimes negative. Their argument is that the weaker coherence requirement, in allowing such bets, treats them as bets with possible outcomes, some of which (the negative ones) are assigned zero probability. It is preferable, they claim, to assume that no possible event can have probability zero.

Operating with weak coherence, as we are, we might then postulate a class of possible events with probabilities zero—events which, while logically possible, are ignorable. Dawid (who doesn't, I should say, accept this argument) suggests the example which I utilized earlier of a dart whose tip is a geometrical point hitting a specified geometrical point on a dartboard as a possible event of zero probability. The obvious worry this raises is of a kind of probabilistic Sorites problem: will the probability of the dart hitting a specified area 10^{-100} times the area of the board also be zero? This worry can be allayed, it is suggested, provided we adopt an account of probabilities based on finite additivity (as a committed follower of de Finetti would); then it remains open for the union of a countably infinite set to have positive probability even when each member of the set has zero probability, provided that that positive probability is a hyperreal smaller than all positive standard reals—an actual infinitesimal.

The calibration result is then treated as an injunction to treat mal-calibration as ignorable but not impossible—to observe what Lindley calls 'Cromwell's rule' (from 'I beseech you in the bowels of Christ to consider that you may be mistaken'), thus defending oneself against the charge of irrational self-confidence.

The cost of such a manœuvre is enormous. With non-standard analysis and non-standard measure theory we admit into our ontology actual infinitesimals when the central concern of modern mathematical analysis has been to eradicate them in favour of theories of limits. It may be, though this is highly controversial, that there are areas of mathematics in which the gains outweigh the costs (see e.g. Robinson [1970] or Hurd [1983]). But, in the present context, we find ourselves faced with *another concept* of necessity, distinct from both logical and natural necessity; we saddle our epistemology with degrees of certainty, i.e. probability 1 less an infinitesimal, probability 1 less twice that infinitesimal, etc. The gain is only that we can coherently proclaim the bare logical possibility of our being less than perfectly calibrated, while still remaining tied to claiming that the probability of that event is zero.

9. A final option, or range of options, is to accept that the calibration theorem does entail a total self-confidence of a specified kind, but to deny that that kind of total confidence is irrational. Van Fraassen in the [1984] paper I have already mentioned takes this line, using arguments closely connected with the calibration idea to support a voluntarist account of partial belief. His main argument I described earlier (2): it leads him to postulate a new requirement of rationality, over and above coherence, which he calls Reflection:

$$p_t(A|p_{t'}(A) = r) = r.$$

(Skyrms, in [1980a], had already advocated a synchronic version of this principle, i.e. a version where $t = t'$). He now gives the calibration theorem in a 'reflective' form: he points out that if you add to a perfectly calibrated set of judgements any judgement that there will be a discrepancy between an actual frequency and your announced forecast—say, 'the probability of rain on days when I announce the probability of rain to be 0.8, equals 0.7'—the enlarged set cannot be well calibrated whatever happens. It would be irrational, given that good calibration is desirable, to organize your beliefs in such a way as to ruin, a priori, the possibility of perfect calibration. So the calibration result, as a consequence of the coherence constraint on rational judgements, simply expresses the irrationality of allowing discrepancy judgements into your probability distribution.

Although there are some errors in the argument leading up to the Reflection principle—in particular, a view of coherence requiring an asymmetry of freedom to bet between an agent and a Dutch bookie, and a view of conditional probability which defines it via what should be a theorem of compound probabilities—both the principle and the cast it gives to calibration are quite acceptable to a strong subjectivist. The errors I mentioned lead only to the mistaken notion that that Reflection is something other than a consequence of coherence. Consequently, van Fraassen is right to see the principle as pointing to partial belief's being a matter of cognitive engagement, of commitment; right, too, to support William James's famous claim that it need not be wrong to believe something on insufficient evidence. The calibration theorem is then explicable as a requirement to stand by one's cognitive engagements, just as by any other commitments. Where he goes wrong is in supposing that this requirement can only be defended within a strongly voluntarist epistemology with all its associated problems about the direct acquisition of beliefs through an exercise of the will. (Indeed, since voluntarism is generally taken to be incompatible with holding that some empirical beliefs are rationally compelled, it seems paradoxical to look to voluntarism to defend the self-confidence requirement embedded in the calibration result.) It can be defended, without such dramatic epistemological shifts, if we can formulate a notion of the rationality of probability-judgements which allows that total confidence in our own calibration is rationally compelled, yet satisfies the intuition that doubts as to its justification are also reasonable and, rightly, common.

None of these responses, then, provides an adequate solution to the calibration paradox: but there are, within (2) and (9), the germs of a solution. To articulate it fully must wait until after (Ch. 2) a subjectivist analysis of convergence and of higher-order probability has been developed.

1.5. COHERENCE: DESCRIPTIVE, NORMATIVE, OR REGULATIVE?

We have arrived, then, at a conception of probabilities as nothing other than the degrees of belief of coherent individuals: individuals who, when forced to bet on any combinations of events at odds representing their degrees of belief, cannot be subject to a Dutch book in any instance. The theorems proved by de Finetti [1937] establish that such degrees of belief can serve as interpretations of the measure 'probability' of the standard calculus—this is the weaker, personalist, claim of de Finetti's theory. Before going on to consider the stronger, subjectivist, claim—that no other interpretations are necessary to account for any uses of 'probable'—we need to explore the issue of how, and to whom, the coherence requirement is to apply. Who are these coherent individuals? Is subjectivism a descriptive theory of actual degrees of belief,

or a normative prescription as to the shape probability-judgements ought
to take?

There is an obvious difficulty for the theory here. If it is construed
as purely descriptive then it is simply false—the beliefs of any actual
individual will exhibit incoherence somewhere. No one is entirely coherent in probability-judgements, just as no one adheres exclusively to
valid arguments. But a purely normative construal of the theory is also
unattractive, in that it drains it of empirical content. Suppose we take
the theory to be simply the following claim: since, if one were perfectly
rational, this is how one's judgements would interrelate, one should, in
pursuit of the evident good of perfect (or at least greater) rationality,
ensure that one's judgements conform (or nearly conform) to the coherence requirement. How could such a normative canon be informative
(as it is claimed that subjectivism is) about the processes whereby the
actual divergent judgements of imperfectly rational agents do, in practice, converge?

In what follows I attempt to pursue a middle way, presenting subjectivism as an idealized descriptive theory which, by virtue of being an
idealization, acquires a normative dimension. But first it is necessary to
meet a challenge stemming from a number of well-known experiments
on actual probability-judgement, carried out (mostly in the 1970s) by
Kahneman and Tversky, Phillips and Edwards, among others. The point
of these experiments is not, as some commentators, for example
Blackburn [1981], have supposed, an attempt to demonstrate that there
is widespread error and confusion in actual probability-assessment. Such
a demonstration would cause problems only for the most naïvely descriptive theory which laid claims to exact, rather than approximate,
fidelity to the facts; but it would no more undermine the subjectivist
claim to provide an idealized descriptive theory than the discovery of
widespread arithmetical incompetence undermines the claim of text-
book arithmetic to being a correct idealized description of our numerical undertakings. Rather, the experimental results are taken to show
that actual probability-judgements deviate systematically, predictably,
and incorrigibly from the judgements to be expected of coherent agents
acting in conformity with the probability calculus. Such a conclusion
would not be too devastating for an objectivist, for whom our degrees
of belief are merely estimates (more or less good) of objective chances.
But the subjectivist seems to be forced into making a choice: either to
take the calculus as somehow given as a norm of rational judgement,
concluding that the experiments demonstrate human irrationality; or,

taking these experimental results as explananda, to conclude that no probability interpretation of the standard calculus can be adequate as an explanatory theory covering all probability-judgements. Either alternative will rule out a unitary conception of subjective probability as idealized descriptive theory with normative content. This challenge must be met before we can give a positive account of the regulative ideal of coherence.

(a) *What is the force of the experimental results?*

The experiments reported in Phillips and Edwards [1966] and Kahneman and Tversky [1972, 1973, 1974, 1982] purport to identify six main fallacies in the reasoning of groups of subjects:

    (i) undue conservatism in revising probability estimates: failure to apply Bayes's theorem
   (ii) insensitivity to prior probabilities, as in the lawyer–engineer example
  (iii) insufficient attention to sample size, as in the hospitals example
  (iv) fallacious expectations of randomness in short sequences
   (v) the 'gambler's fallacy'; expecting independent event-sequences to even up in conformity with a proverbial, and quite mistaken, 'law of averages'
  (vi) an 'illusion of validity': overconfidence in one's own appraisals.

It is by no means obvious that these reported results are trustworthy in the sense of being experimentally well established. There are, in at least some of these areas, conflicting experimental findings; and the tendency claimed to be observed as (i) is hard to reconcile with that observed as (ii). Unless one is seeking an explanation for the re-election of a Tory government despite exit polls to the contrary, it cannot be plausible to accept that people in general are both systematically over-conservative and not conservative enough.

There are also at least two aspects of the experimenters' methodology which ought to arouse disquiet. Kahneman and Tversky, in particular, treat Bayesian decision theory as a preconceived norm, the origins of which need not be inquired into. But it is clear that, once we do ask where these theories originate, we can no longer equate the competence of the experimental subjects with their observed performance. As Cohen [1981]—with the support of some commentators—argues, any theory of applied logic or everyday probabilistic reasoning which is at least in part normative must be based ultimately on the data of untutored intuitions in that area; hence it must ascribe to ordinary agents a cognitive competence to reason in accord with the theory. There is no alternative basis, since there is no means of supplying either an empirical or a metamathematical justification of the application of a deductive logic or an interpretation of the probability calculus to everyday reasoning (as distinct from a demonstration of their internal soundness). Without such an independent justification there is no escaping, ultimately, an appeal to intuition as to people's competence: if their performance is, as it usually is, observably inferior to their competence then this must be explained in terms of cognitive illusion; environmental factors, defects of education, intelligence, or motivation—but not in terms of human irrationality. To make the latter claim would be to apply a theory derived from a set of intuitions to argue the faultiness of those intuitions— a process which amounts to sawing away the branch one is sitting on (see Kahneman and Tversky [1979], Cohen [1979], and, for a refined version of his earlier views, Cohen [1991]). It is the inanity of such a process which leads Kahneman and Tversky into attempting to explain agents' judgements in terms which simply redescribe their judgements (see Gigerenzer *et al.* [1989]).

Moreover, the experimenters attempt to establish the alleged fallacies by reference, in most cases, to information about relative frequency; subjects who ignore relative frequencies are then taken to be behaving irrationally without the need for any qualification as to their competence. But this experimental method is radically incoherent. Subjects are asked to make personal assessments of probability without prompting as to the nature of those assessments (if they are prompted to think of probability in a frequentist way the 'fallacious' behaviour tends to disappear) and it is then assumed that their responses, which represent degrees of belief, ought simply to embody frequencies. But any defensible personalist account of probability rejects such a simplistic reliance on frequency; if I am asked to judge the probability that a particular man described to me is a lawyer I can take into account not only anything I know about relative frequencies but also anything I believe as to which frequencies might be relevant, the necessity for coherence with my other beliefs, and the extent of my knowledge—and I may do all this on a quite different basis from the way in which I would act if asked to judge a long run of events. Kahneman and Tversky are assuming that their subjects are naïve frequentists, compelling them to make judgements of probability which, being singular, would be regarded as inadmissible on most frequency interpretations of probability and then concluding that, where their answers differ from the long-run established norm, their behaviour is irrational.

We would do well, then, not to place too much reliance on these experiments. Still, perhaps better-designed and methodologically less dubious experiments might be devised which would arrive at similar results. Would that force upon us the descriptive–normative dilemma I outlined earlier? I believe that, granted only a competence–performance distinction, all these results can be accounted for within the framework of a subjectivist theory of probability linked to the standard formal calculus. (They can be accounted for in other ways too: Cohen's Baconian analysis, developed throughout the eighties; Lopes and Oden's [1991] connectionist arguments; Adler's [1991] development of an Austinian account of conversational implicatures.)

Let us consider each of the six 'fallacies' in turn.

(i)

The first is exemplified by a well-known experiment carried out by Ward Edwards and colleagues in the 1960s. To quote the version described by Peterson and Beach [1967: 32], subjects were posed this problem:

Two urns are filled with a large number of poker chips. The first urn contains 70% red chips and 30% blue. The second contains 70% blue chips and 30% red. The experimenter flips a fair coin to select one of the two urns, so the prior probability for each urn is .50. He then draws a succession of chips from the selected urn. Suppose the sample contains eight red and four blue chips. What is your revised probability that the selected urn is the predominantly red one?

Most subjects judged that the data raise the probability that the urn is the predominantly red one from 0.50 to about 0.75. Bayes's theorem, however, gives a posterior probability of 0.97. So, over-conservatism in revising prior probabilities appears to be rife.

However, the experimenters here are assuming both that one can assign prior probabilities directly in terms of given ratios and that posterior probabilities should be arrived at on the basis that the set-up is a chance one. But the first assumption ignores the subject's entitlement to take into account the reliability of the witness telling him or her the relevant ratios (since this is in such cases the experimenter, subjects will normally place that reliability very high, but not necessarily at 100%); and the second assumption ignores the reasonableness for the subject of being rather cautious as to assuming the arrangement of chips within the bags to be pure chance—caution which becomes more reasonable as the task becomes more important. So, rather than Bayes's theorem not being applied, we can interpret the subjects' behaviour as consistent application of the theorem but with quite rational assumptions, different from those of the experimenters, about the prior and conditional probabilities appropriate to the set-up.

(ii)

The second case is exemplified by the 'lawyer–engineer' problem, described by Kahneman and Tversky [1973: 241]. One group of subjects was given the following task:

A panel of psychologists has interviewed and administered personality tests to 30 engineers and 70 lawyers, all successful in their respective fields. On the basis of this information, thumbnail descriptions of the 30 engineers and 70 lawyers have been written. You will find on your form five descriptions, chosen at random from the 100 available descriptions. For each description, please indicate your probability that the person described is an engineer, on a scale from 0 to 100.

A second group was given the same instructions, except that they were told there were 70 engineers and 30 lawyers. Both groups were given the same descriptions, of which this is typical:

Jack is a 45-year-old man. . . . He is generally conservative, careful and ambitious. He shows no interest in political and social issues and spends most of his free time on his many hobbies, which include home carpentry, sailing and mathematical puzzles.

For each description, the mean probability-judgements in the two groups were about the same. From this it was inferred that base rates were being largely ignored by the subjects, and that this neglect was a systematic bias in their reasoning.

But here, in fact, one of the experimental results points us towards a different inference. Kahneman and Tversky [1973: 242] report that 'people respond differently when given no specific evidence and when given worthless evidence. When no specific evidence is given, the prior probabilities are properly utilised; when worthless evidence is given, prior probabilities are ignored.' Stripped of its objectivist preconceptions, this is merely to say that, if we have no particular information at all, we tend to assign probabilities in line with relative frequencies; but that when we do have some information we try to extract a basis for our judgements from it, no matter how scanty it is. Which may show that experimental subjects assume too readily that the information given is relevant; or that it is not as irrelevant as the experimenters suppose; or that, because we prefer where possible to reason from causes rather than chances, we are over-disposed to conclude that we have found causal links. But these are only factors affecting our performance, not our competence to reason in a Bayesian way.

(iii)

The third case, that of insufficient attention to sample size, is exemplified by such behaviour as that of generalizing from the male : female ratio in a small number of births in one hospital to expecting the same male : female birth ratio in a larger hospital. But here, as Cohen [1982] points out, this behaviour can be explained not only as an attempt to apply Baconian measures but also, in entirely Pascalian terms, as a simple lack of statistical knowledge on the subjects' part affecting their performance. It can even be shown experimentally that 'subjects can be led to acknowledge and apply the law of large numbers about sample-size by a procedure which suggests that the law was already in their competence as a piece of tacit knowledge' [1982: 268].

(iv)

The fourth case is where, Kahneman and Tversky [1972: 432] claim, people fallaciously expect the essential characteristics of a random sequence to be exhibited even in short sequences—what they pejoratively term belief in a 'law of small numbers'. For example, consider the response to subjects of the following problem:

All families of six children in a city were surveyed. In 72 families the exact order of births of boys and girls was GBGBBG. What is your estimate of the number of families surveyed in which the exact order of births was BGBBBB?

Most subjects gave an answer less than what the experimenters take to be the correct answer—72. But in fact, for a subjectivist, there is nothing irrational in the subjects' expectations, even if in general the conditions which would vindicate them are not fulfilled. We can see this from a trivial theorem of de Finetti's [1974: i. 202] which Jeffrey [1984] dubs (non-pejoratively) 'de Finetti's law of small numbers': in the case of coin-tossing, if we are convinced that the relative long-run frequency of heads is $\frac{1}{2}$ and that any probability-measure which adequately reflects our state of belief should not vary from trial to trial, then our subjective probability = success rate = expected frequency of heads.

But these two conditions of which we must be convinced for this to be true derive from our experience of finite, usually short, sequences. If we are presented with a sequence where the actual frequency of heads is not $\frac{1}{2}$ and where there is asymmetry, we may quite reasonably take such a sequence to be less random if we take 'random' to mean that we are entitled to hold these two convictions.

(v)

The fifth case is that of the well-known 'gambler's fallacy', where subjects appear to be convinced that, for example, a long run of red on the roulette wheel makes black better than an even-money bet on the next spin. Even here, several answers are possible other than taking such judgements to be systematically irrational. First, we might claim, as for (iii), that this fallacy simply shows a widespread and correctable ignorance of statistics, rather than impugning our competence. Or, secondly, we might note that the experimenters' instructions suffer from an overarching ambiguity. Are the subjects being asked to judge the probability of a single outcome in a sequence where they would be justified in holding the two convictions mentioned in (iv)? Or are they being asked instead to judge the relative probability of at least one tails outcome in n tosses to that of an all-heads sequence? This ambiguity cannot readily be removed: for to explain clearly to the subjects what is intended would be to instruct them sufficiently to discourage any incorrect judgements.

The final case gives rise to the claim (Kahneman and Tversky [1974]) that, by a similar heuristic to that of representativeness which they see as accounting for (ii) and (iv), many people experience unwarranted confidence in their own highly fallible judgements—they suffer from an illusion of validity. One can explain this phenomenon in terms of a calibration effect. Given only that our probability distribution is coherent, we must assign probability 1, relative to that distribution, to our being well calibrated (i.e. our estimates corresponding to our experience). The solution here is the solution to the calibration problem, to be presented in Chapter 2.

It is certainly arguable—and it is a very important argument—that even though these experimental results are not compelling they do suggest that in some contexts judgements which have every right to be called probability-judgements might not be attempts to apply the standard probability calculus. We must, it may be claimed, distinguish situations in which agents try to judge in conformity with the Pascalian calculus from those in which they attempt to apply a non-complementational concept; or situations where counterfactualizable judgements are being attempted from those where agents are concerned, non-counterfactualizably, with relative frequency within a particular class.

If such claims were correct, they would certainly restrict the scope of the subjectivist theory which I am offering; but they would not challenge its applicability within some limits. I give some illustrations of this approach in Chapter 4: I want here only to argue that despite these experimental results actual judgements can be treated as the judgements of agents whose competence is that of coherent agents, as the theory presents that notion, but whose performance is affected by their practical limitations. The way is clear then to consider how descriptive and normative aspects of the theory coexist.

(b) *Coherence as a Regulative Ideal*

In their early formulations of subjectivism both Ramsey and de Finetti stressed a view of the theory as an idealized description. Ramsey compared it with Newtonian mechanics, which is none the less useful for being known to be, strictly, false; de Finetti presented it as an idealized analysis of our modes of thought and behaviour under uncertainty. When it came to accounting for the character of such idealization, however, both allowed themselves to be drawn into a damaging psychologism. Ramsey [1931: 174] postulated a 'law of psychology that . . . behaviour is governed by what is called the mathematical expectation'; de Finetti [1937: 152] claimed airily that there are 'rather profound psychological reasons which make the exact or approximate agreement that is observed between the opinions of different individuals very natural'. Such psychological laws, we must suppose, would contain within them the justification of this idealized theory, permitting us to treat its known lack of strict fidelity to fact as inessential. But neither de Finetti nor Ramsey ever tried to develop this part of their theory so as to support such a claim: had they attempted to do so, the relegation of probability theory to psychology would have threatened the whole enterprise of developing a logic of partial belief.

A preferable account of the idealization inherent in the theory can be based on the distinction, introduced in the previous section, between competence and performance. Deductive logic, applied arithmetic, and subjective probability theory should all be seen as codifications of human competence in a particular field of reasoning. Extracting a description of such competence involves ignoring performance errors and the factors which create them; this deliberate simplification (which necessitates a known departure from the actual) is idealization in just the same sense as occurs in, say, the kinetic theory of gases. The resulting theories comprise sets of rules which are systematic summaries of what is essential to a particular practice, in just the same way as rules of grammar are extracted from linguistic practices. No actual individual need be supposed to act always, or be capable of acting always, in accord with all these rules; but the ultimate ground of the rules lies in actual reasoning behaviour.

In its functioning, probability theory is, then, better described as regulative rather than normative: it constrains our actions rather as rules of grammar constrain linguistic practice than as moral precepts constrain conduct. This thought lies behind de Finetti's ill-formulated efforts to present coherence as a thoroughly permissive constraint which does, however, impose a normative demand: the air of paradox is dispelled, he claims, when one realizes that the demand is never of the form 'You must do this' but only 'You must do this if you wish to avoid certain consequences'. Coherence has regulative force because it arises in characterizing competence. Unlike moral precepts, it does not purport to govern our ends—it does not insist that certain combinations of judgements absolutely could not be permitted. Rather, just as grammar constrains how we can speak if we aim to exercise a certain linguistic competence, so coherence constrains the judgements we can make if we are to exercise competence in probabilistic reasoning, the competence codified by the idealized descriptive theory: it governs the means to that end.

Coherence is a regulative ideal of the same nature as deductive consistency, as we noted earlier. Both derive their regulative force because, when we codify certain areas of actual untutored reasoning, we find that they appear as regularities in our practice as we strive for certain ends. To seek any other normative basis for either would be mistaken. It might well be useful, for instance, to codify certain reasoning practices where our aim was, not to avoid all contradictions, but to ensure that local contradictions did not become global; in doing so, paraconsistency rather than consistency would emerge as the constraint on our judgements, given that aim.

Similarly, some constraint other than coherence is perfectly possible if we are dealing with probabilistic reasoning not necessarily governed by the standard mathematical calculus. Coherence derives its particular force from being that constraint upon our judgements necessary and sufficient to ensure that they fulfil the aim of complying with this calculus.

2

Convergence and Consensus

THIS chapter gives an account of the dynamics of SCS.

§2.1. presents and criticizes the standard view of belief change via Bayesian conditionalization and argues the advantages of defining 'conditional' probability by means of pay-off tables.

§2.2. presents the de Finettian concept of exchangeability, discusses its role in the convergence of probability-judgements onto observed frequencies and asks whether it is an objective property of the world or a property of our subjective judgements.

§2.3. examines how conditionalization and exchangeability contribute to ensuring consensus about a large range of probability-assessments.

§2.4. tackles the criticism which can be launched at any coherentist theory that it fails to take account of weight of evidence and offers a characterization of weight in terms of resiliency and second-order probability.

2.1. BAYESIAN CONDITIONALIZATION

(a) *The Standard View*

For strong subjectivism, the only constraint of rationality on our judgements is that they should be coherent. Such a constraint, as we have seen, is to be thought of as applying to the partial beliefs—monadic or conditional—of some (ideal) individual at some particular time; beliefs held at time t_1 need not be coherent with those held at time t_0—such a requirement would render rational change in our beliefs impossible. But subjectivism cannot, if it is to provide a complete interpretation of probability, rest content with defining the limits of the agent's partial beliefs at some instant. We need some account of how beliefs may be altered by experience; and, even without the passage of time, we are interested in delimiting the probability-values a rational agent may assign not just to the outcomes of events considered in isolation but to the outcome of an event given that various other events have or have not occurred. The challenge for subjectivism is to offer an account of posterior and conditional probabilities which illuminates the connection between them without imposing upon probability-judgement any further constraint than coherence.

The subjectivist answer rests upon the process of 'conditionalization'. As I pointed out in §1.3, weak coherence—interpreted as immunity from Dutch books—can be defined for conditional probabilities and between monadic and conditional probabilities. Suppose that at time t_0 you are coherent in your evaluations of (monadic) probabilities $p(E)$ for all events E in your universe of discourse. It can be shown (de Finetti [1974: i. 136–9]) that you will be coherent in your conditional evaluations $p(E_i|E_j)$ if the values you assign to all your $p(E_i|E_j)$ comply with the relation $p(E_i|E_j) = p(E_i \wedge E_j)/p(E_j)$ or its corollary, Bayes's theorem. Now suppose that at time t_1 you come to know for certain the outcome of an event E_k about which you were, at t_0, uncertain. If you now alter your monadic probability distribution across all events E so that your new $p(E_i)$ is equal to the previous $p(E_i|E_k)$, it is clear that your new distribution will be coherent iff you make this change in compliance with Bayes's theorem. This process is known to the subjectivist as conditionalization—and, to mark the central role of Bayes's theorem, as Bayesian conditionalization.

Here, the subjectivist claims, is a prescription for rational belief-change which does not commit one to adopting any constraints over and above coherence: as we shall see in the next two sections, adding one other ingredient to the prescription will, it is claimed, enable it to dissolve the problem of explaining convergence of judgement. It is a strikingly simple prescription. Any events which are sufficiently well defined for us to be able to discriminate their outcomes can be conditionalized over as we acquire new knowledge. The complex distinctions elaborated in other theories between inductive and statistical or epistemic probabilities are jettisoned. Diachronic rationality, we might say, becomes merely a matter of exercising the same skills required for synchronic rationality: if you can distribute your conditional probabilities at t_0 so as to be coherent then, all you need do at t_1 is to adjust your prior monadic probabilities to the appropriate subset (given new information) of the t_0 conditional probabilities. It will be important to keep constantly in mind, in what follows, the possibility that this prescription is too simple to do justice to the complexity of conditional belief. We should emulate Ramsey's caution (though not his apparent [1931: 180] transposition error!):

'the degree of belief in p given q' . . . does not mean the degree of belief in 'If p then q', or that in 'p entails q', or that which the subject would have in p if he knew q, or that which he ought to have. It roughly expresses the odds at which he would now bet on p, the bet only to be valid if q is true. . . . This is not the same as the degree to which he would believe p, if he knew q for certain . . .

(b) *Difficulties for the Standard View*

There are at least five difficulties in this concept of conditionalization which need to be met: the first two can be dealt with, I believe, without strain within the standard theory; the third and fourth, however, pose genuine problems which, I argue, can best be met by radically altering our view of the meaning of conditionalization; the last suggests a limitation on what should be expected of subjectivism.

(*i*)

Many writers, e.g. Hacking [1967], Ellis [1973], Kyburg [1978*a*], have seen conditionalization as incompatible with a pure coherentist subjectivism. They claim that conditionalization amounts to a regulative principle of rationality which stands in need of, but cannot possess, a justification within the subjectivist system. They point, for example, to such formulations of the concept as that of Skyrms [1975: 145]:

to move to a 'final' set of epistemic probabilities which accommodate a new item of knowledge in a rational manner (adopt) Rule C: for any statement take its new probability to be its old probability conditional on the new item of knowledge.

What is the force of this rule? Is not a subjectivist compelled to admit that I am behaving rationally provided that my initial and final set of probabilities are both coherent, no matter how I have moved from one position to the other?

This criticism only works if coherence is taken to be limited to monadic probabilities at some particular time: it then follows, indeed, that the coherence constraint determines what may be in the picture if a series of snapshots of our beliefs is taken—what happens in the gaps between remaining unconstrained. I argued earlier that coherence can and should be extended to conditional probabilities: you are coherent at any time iff there is no subset of your conditional beliefs (not only your monadic judgements) which could be chosen so as to force you into a series of bets resulting in certain loss. When you acquire a new item of information your new monadic probability distribution will, in general, conflict with your prior monadic distribution—but it *must* be coherent with your prior distribution of probabilities conditional upon the truth of this new item, which, as we have seen, could only have been themselves internally coherent if they complied with the theorem of compound probabilities. This dyadic–monadic coherence is thus a temporally extended constraint (whereas monadic and dyadic coherence are instantaneous constraints) and the rule of conditionalization is no more than a direct corollary of the coherence requirement. No new limitation on our judgements is entailed by the process of conditionalization: as de Finetti says [1974: i. 140],

> assuming coherence, conditional probability is not a new concept; the acquisition of a further piece of information . . . acts always and only in the way we have just described—suppressing the alternatives that turn out to be no longer possible.

(*ii*)

Since conditional probabilities $p(E_i|E_j)$ must, to be coherent, equal $p(E_i \wedge E_j)/p(E_j)$ it might be thought that conditionalization depended on a multiplication theorem enabling us to establish $p(E_i \wedge E_j)$, and that that in turn depends upon our distinguishing independent events from others. And, as is well known, the concept of independence is difficult to define in any non-objectivist manner.

This problem is only an apparent one, however. All that is necessary to conditionalization is that we possess some rule for assigning probability-values to $p(E_i \wedge E_j)$ for every i and j—it could just as well be a deviant rule as the standard product rule: that would simply provide us with an interpretation of a deviant probability calculus. Granted the standard calculus, we are committed to a standard multiplication theorem, but we do not need to be able to define independence in order to apply such a rule—as we shall see in §2.2, exchangeability will do equally well.

(*iii*)

A much more serious problem for the standard theory of conditionalization is the confusion surrounding the subjectivist interpretation of the probabilities $p(E_i|E_j)$ and their betting-measures. As we have seen, $p(E_i|E_j)$ is defined by means of the theorem of compound probabilities:

hence it is a measure of the degree of our coherent belief in the occurrence of both E_i and E_j given the occurrence of E_j. Subjectivists (though
not only subjectivists) have tended to assume, in terming $p(E_i|E_j)$ a
'conditional probability', that $p(E_i|E_j)$ will necessarily equal $p(E_j \to E_i)$.
But as Cohen [1970] and Lewis [1976] point out, whether $\to$ is construed truth-functionally or not, we cannot thus equate 'conditional
probability' with the probability of a conditional. Lewis demonstrates
that doing so will lead to the triviality result that $p(E_i|E_j)$ must al-
ways equal $p(E_i)$; and Cohen shows that, since $p(E_j \to E_i)$ must equal
$p(\neg E_i \to \neg E_j)$, but we can cite instances where $p(E_i|E_j)$ is not equal to
$p(\neg E_j| \neg E_i)$, we cannot identify $p(E_i|E_j)$ with $p(E_j \to E_i)$. This distinction produces an ambiguity in the notion of a conditional betting-
quotient as a measure of $p(E_i|E_j)$. Does this quotient represent the odds
I should accept on $p(E_i)$ if I am told that I will be taken up on that bet
if E_j is true but the bet will be called off if E_j is false? Or does it
represent a measure of the degree of belief in E_i which full belief in E_j
would currently warrant me in having? It is far from obvious that these
quotients should be equated. I shall explore a possible solution to this
problem in (c) following.

(*iv*)

Another serious problem for the standard theory is the restriction of
conditionalization to cases where the newly acquired evidence is to be
taken as certain. We need not even be convinced by empiricist arguments that certainty is never warranted for the contingent statements
which typically express the results of novel observations to find this
restriction worrying: it would be irrational in the extreme to deny that
we ought on some occasions at least to change our beliefs under the
pressure of new evidence even if we would not claim to be certain of
the truth of such evidence. If conditionalization is to comprehend all or
most belief-changes, we must, it seems, either offer an account of evidence which warrants taking for certain in one context what is not claimed
as certain in other contexts or modify the rule of conditionalization
being used to incorporate the probability of the relevant evidential statements. The former option disturbs the neat de Finettian dichotomy
between the realms of certainty and uncertainty which underpins his
distinction between prevision and prediction; the latter option yields
rules of conditionalization which seem to go beyond the coherence
constraint. I take up this problem in (*d*) following.

(v)

One very large issue deserves mention here. Subjectivists generally make very large claims for conditionalization as a recipe for belief-change in the light of evidence: it eliminates any need to offer differing interpretations of inductive and epistemic probabilities (or, if we like, conditionalized monadic probabilities are the 'inductive' probabilities derived from prior monadic probabilities which are 'epistemic') and it dissolves the problem of induction—thus, de Finetti [1974: ii. 202] 'in our formulation the problem of induction is no longer a problem ... Everything reduces to the notion of conditional probability.' Such a position can only be sustained if the large range of arguments against identifying inductive support with conditional probability-relations and against reducing confirmation to increase in probability can be met. If they cannot, however, it does not follow that we should abandon conditionalization as a model for belief-change—merely the general claim that it models all the kinds of belief-change we experience. Perhaps, for example, conditionalization accurately represents the way in which we should modify the probability-values attached to judgements about single events or exchangeable collections of events, but some other relation governs the support one generalization can lend another or the support a body of evidence can lend to a generalization. I return to this issue in §4.1 and §4.2.

(c) *Another View of Conditionalization*

The standard view, as we have seen, *defines* $p(E_i|E_j)$ via the relationship $p(E_i|E_j) = p(E_i \wedge E_j)/p(E_j)$; coherent conditional betting must meet the same constraint. But there is another approach to conditional probability, through what de Finetti [1972] calls the 'approach through losses', which suggests itself to us once we recall that coherence ought to be defined as immunity from Dutch book, with conformity to the calculus as a consequence of that rather than the reverse. This approach was first outlined by Skyrms [1977]; I here—in common with Howson and Urbach [1989]—make use of his treatment (which I have also utilized [1987]).

Let us begin by defining a conditional bet in terms of the pay-off table for such a bet. A bet on Q given P at odds b/a is defined as the bet the pay-off table for which is Table 1. Now suppose we look for a combination of simple bets on P and Q which result in this pay-off table. One way (though not the only way) of doing this is by a 'hedging'

TABLE 1

P	Q	Pay-off
T	T	a
T	F	$-b$
F	T	0
F	F	0

TABLE 2

P	Q	P&Q	Bet 1	¬ P	Bet 2	Sum of bets 1&2
T	T	T	c	F	$-f$	$c-f$
T	F	F	$-d$	F	$-f$	$-d-f$
F	T	F	$-d$	T	e	$e-d$
F	F	F	$-d$	T	e	$e-d$

TABLE 3

P	Q	Pay-off
T	T	$c-f$
T	F	$-(d+f)$
F	T	0
F	F	0

strategy. Consider the sum of two bets: a bet at odds c/d on $P\&Q$, and a bet at odds e/f against P(i.e. on ¬ P). This yields the pay-off table shown as Table 2. Clearly, if we choose the odds so that our winnings on the second bet equal our losses on the first—that is, $e - d = 0$—this pay-off table will be the same as for a bet on Q given P at odds $(c - f)/(d + f)$, which gives Table 3. How can we now move from conditional bets to conditional probabilities?

Recall that the probability of any event is the betting-quotient you must choose in order to avoid any subset of bets resulting in certain loss—in order, more simply, to yield fair bets, ones with an expected value of zero. This last betting arrangement is fair if:

$\{c - f\} \{p(P\&Q)\} - \{d + f\} \{p(P\& \neg Q)\} = 0$ so that
$$p(P\&Q)/p(P\& \neg Q) = (d + f)/(c - f).$$

For any value of p(P), other than zero, the bet will be fair iff this value is partitioned between p(P&Q) and $p(P\& \neg Q)$ in the ratio $(d + f)/(c - f)$. So,

$$p(P\&Q) = \{(d + f)/[(d + f) + (c - f)]\} \{p(P)\}$$
$$= \{(d + f)/(c + d)\} \{p(P)\}.$$
Then $\{p(P\&Q)\}/\{p(P)\} = (d + f)/(c + d)$.

We can now define the conditional probability of Q given P as the betting-quotient $(d + f)/(c + d)$ which you must choose so that a compound bet at odds c/d on P&Q and d/f against P cannot certainly result in a loss for you. The relation of compound to conditional probability is now an elementary theorem of the interpreted calculus, *not* part of the definition of Q|P: the argument above is its proof.

This approach enables us to deal with the difficulties I raised in the previous section.

Notice, first, that we have developed this definition of $p(Q|P)$ without making use of the notion of implication. We have a well-defined decision procedure for bets on Q given P which depends only on the ability to sum the pay-offs of simple bets. Suppose someone insists on knowing the probability of 'if P then Q'. You should begin by demanding clarification of how this conditional is to be understood between you, and whether it can be represented by formal or material implication. If no such clarification is possible then you assign no meaning to p (if P then Q)—the event is not well defined. If you and your opponent do agree, then you will define the bet on 'if P then Q' by means of a pay-off table which might, but need not, take the form of the table we have seen for $Q|P$. You might agree, for instance, that anyone betting that 'if it rains today it will be foggy tomorrow' will win his or her bet in every case except that of rain today and no fog tomorrow, so that the pay-off table would be as in Table 4. In this case the probability of the conditional sentence is the value of b/a you should choose for this bet in order to avoid a Dutch book; it will not, in general, be the same as the value you would choose to bet on fog tomorrow given rain today, and it is quite feasible that you would choose a different value if the conditional is to be defined by a different pay-off table. It may well be possible to assign probabilities to conditionals (§4.3 will pursue this point); but what is generally termed 'conditional probability' is a different type of judgement

TABLE 4

Rain today	Fog tomorrow	Pay-off
T	T	*a*
T	F	−*b*
F	T	*a*
F	F	*a*

which might, if the terminology were not so entrenched, better be referred to as the dyadic probability of *Q* on *P*, or the probability of *Q* relative to *P*—a matter of choice of a betting-quotient when certain alternatives (bets where not-*P*) are eliminated.

Finally, we should note that the mechanism of conditionalization will remain unaffected by this redefinition of conditional probability: the process still remains that of adjusting prior values to the value of *Q|P* once *P* is known. Moreover, it becomes much easier to justify the compulsory nature of conditionalization as a consequence of coherence if we do so through the loss approach. Skyrms [1975] offers a very simple proof (derived from de Finetti's more general proof [1972]) that iff beliefs are changed by conditionalization will fair bets conditional on *P* yield fair monadic bets.

(d) *Evidence and Certainty*

If conditionalization is to be at all plausible as an account of judgement it must cater for cases where our beliefs are altered as a result of observations about which we would not wish to claim to be certain. We have formulated the rule of conditionalization as requiring that monadic probabilities $p(E_i)$ should be altered to the values we had previously held for the dyadic probabilities $p(E_i|E_j)$ once we come to know the truth of E_j. What if we strongly believe, but would not claim to know, E_j? One might try to deal with this issue by distinguishing belief from acceptance. The new observations, we might suppose, lead us to assign a very high value to $p(E_j)$, but not to claim that $p(E_j) = 1$; we accept, that is, that there is still some risk in asserting E_j, and that such risk can be assessed by betting-quotient. For the purpose of altering our probabilities, however, we do not need to utilize this value of $p(E_j)$; we will not indefensibly be claiming certainty in our judgements of other E_i if we accept E_j as certain for the purposes of our calculations, since the totality of our monadic and dyadic judgements still must contain a judgement about E_j, on which we are willing to bet, and which thus does not allow us to escape the risk that we are mistaken about E_j.

But there are several problems in taking such a line for the strong subjectivist position. First, a rational agent must accept the need to justify treating some observational results as grounds for acceptance of E_j, others not. And such a demand for justification is incompatible with the subjectivist position that any value of $p(E_j)$ which an agent can assign coherently with his or her other beliefs is acceptable: for then the results of conditionalization would be different for different agents, and the mechanism would fail to account for convergence of beliefs as evidence is acquired. And, secondly, the need to justify treating some observations as evidence seems to demand adherence to rules of acceptance (see Kyburg [1975] and [1978b]) which import objective constraints on our judgements into subjective probability.

Both these criticisms can, I think, be answered. To answer the first will require an account of consensus which will explain how the single constraint of coherence sufficiently limits the probabilities we can assign to E_j posterior upon the other evidence available to us so as to ensure that agents' differing conditionalizations tend to converge—though they will not necessarily come, in a finite time, to coincide. The second can be answered by conceding that acceptance rules may be needed, but treating them as purely probabilistic and hence subjective. I discuss these issues further in §2.3.

A quite different approach to uncertain evidence rests upon taking an observation simply as something which raises degree of belief in E_j, say from $p_0(E_j)$ to $p_1(E_j)$, without raising it all the way to 1. Such uncertain evidence should then affect our beliefs in the same way as conditionalization upon certainties, except that we should amend our probability-values by applying a generalized version of the rule of conditionalization, Jeffrey's Rule [1965]:

If between time t_0 and time t_1 your degree of belief in E_j has altered from $p_0(E_j)$ to $p_1(E_j)$ you will remain coherent iff you alter all your $p_0(E_i)$ to $p_1(E_i)$ such that

$$p_1(E_i) = \{p_1(E_j)\} \{p_0(E_i|E_j)\} + \{p_1 (\neg E_j)\} \{p_0(E_i \mid \neg E_j)\}.$$

This rule, as might be expected, enjoins us to conditionalize on the basis of uncertain evidence by, in effect, a weighted average of conditionalizing on E_j and on not-E_j, deriving the weights from the ratio of $p_i(E_j)$ to $p_1(\neg E_j)$.

Some such rule, it seems, is important to any subjective account of conditionalization which is to have any pretensions to generality. There seem, however, to be two difficulties about incorporating it within strong subjectivism.

First, though we have a proof that coherence will be maintained iff the less generalized conditionalization rule is followed, we do not, so far as I know, have such a proof for the general rule. This proof ought not to be difficult, but will not be sought after here. Secondly, it has been argued that once such a rule is adopted situations will emerge where objective probabilities can and should be distinguished from subjective degrees of belief. But, as I argued earlier, the possibility of transposing the formula does not lead to objective probabilities. What is needed to render probability objective is that rationality should demand that we prefer one distribution to another; nothing in Jeffrey's Rule has any such effect.

To sum up this section: there are two approaches (not mutually exclusive) to dealing with the problem of uncertain evidence. We might develop an account of how coherence leads towards probabilistic acceptance-rules for treating evidence as, in some contexts, certain. Or we might generalize the conditionalization rule to allow for probable evidence increasing other probabilities: and we can accomplish that without fear of being driven to admit objective probabilities. In what follows I adopt the latter approach.

2.2. EXCHANGEABILITY

(a) *Scope*

Most non-personalist interpretations of probability rely upon some notion of symmetry in their definitions of 'probable': classical theories need a concept of equipossibility; a priori theories either use the Principle of Indifference (Keynes) or concede a sense of probability in which it is equivalent to relative frequency (Carnap–early); frequency theories need a concept of randomness (von Mises) or at least of repeatable events (Reichenbach). (An exception is propensity theory, in some of its versions—which explains perhaps why some writers look to the possible combination of that theory with personalism.) As is well known, giving symmetry so central a role produces problems for all these theories: how to define equipossibility other than circularly by reference to probabilities; how to determine the space of alternatives towards which we should be indifferent; how to identify randomness within finite sequences; and what variations are irrelevant when treating an event as 'repeatable'. And, if probabilities are to depend upon symmetry, it becomes a problem whether, and how, to give a sense to talk of the probability of a singular event.

Subjectivism avoids these problems by defining probability without reference to generic events, collections, or reference classes. The probability of occurrence of any event is the degree of belief an ideally rational agent could have as to its occurrence. Any agent is free to form probability-judgements about events which are not repeatable or which possess irreducible differences from all other events: it is legitimate and meaningful to assert that the probability that the British electorate would, by a simple majority, support the Maastricht agreement in a referendum is 0.6, even though we cannot readily make judgements of frequency in some class of similar events. If coherence and conditionalization permit, your personal probability that the next toss of this coin will yield heads may be 0.4 even if, for the last 100 tosses, it has been 0.9.

Of course, most of us, in contexts such as games of chance or actuarial predictions, are in fact influenced very strongly in our probability-judgement by symmetry considerations. The subjectivist does not deny that; instead, (s)he characterizes such events as 'exchangeable' (de Finetti's term) and accepts the onus of accounting in subjectivist terms for the apparently objective statistical laws which govern the probabilities of such events. But, unlike in objectivist theories, symmetry plays no part in the meaning of 'probable': it is, rather, a consideration which bears upon the assessment of some, but not all, probability-values.

(b) *Definition and Fundamental Theorems*

Exchangeability is defined as follows (de Finetti [1974: ii. 68]):

Given a probability space $(\Omega, Å, P)$ and a sequence of events in $Å$: $A_1, A_2, \ldots$, then $A_1, A_2, \ldots$ are exchangeable iff for all subsequences $A_{r(1)}, \ldots A_{r(k)}$, the probability $p(A_{r(1)} \wedge A_{r(2)} \ldots \wedge A_{r(k)})$ depends only on k and not on the values $r(1) \ldots r(k)$.

Less formally, as de Finetti puts it in his [1972: 189]:

A collection of events is exchangeable iff the probability that h of them occur depends only on h and not on the events chosen.

Two interesting features of this definition have been pointed out by Dawid [1977, 1985]. First, exchangeability is referred to a given probability space. That opens up the possibility that a sequence of events in two or more spaces may be exchangeable in some but not all of the spaces. For example, actuaries may create two such spaces by classifying the mortality risks of two individuals according to two different factors, with a sequence of events being exchangeable in one space but not the other. Such partial exchangeability can be a useful concept for the statistician, permitting a wider range of symmetries than would objectively construed independence. Secondly, although as a matter of historical fact the motivation for developing the concept has been to provide the theory of subjectivism with an account of the role of frequencies in probability-judgements (as we shall see later), the definition does not rule out objective construal of probability. So, objectivist theories might plausibly seek to use this concept, rather than the more commital notion of independence, in defining randomness—provided, of course, they did not seek, as many do, to define probability by reference to random sequences. Objectivist hostility towards the subjectivist uses of the concept need not extend to the concept itself.

Turning now to these uses, there are two fundamental theorems, first established by de Finetti, which are often confused by critics of the theory.

(i)

The relationship between exchangeability and independence is clarified by de Finetti's representation theorem:

Let $A_1, A_2, \ldots$ be an infinite sequence of exchangeable events in a probability space $(\Omega, \text{Å}, P)$ and let $\alpha_0 = 1$, $\alpha_k = p(A_1 \text{ \& } A_2 \text{ \& } \ldots A_k)$. Then there exists a unique distribution function $F(p)$ on $[0, 1]$ such that $\alpha_k = \int p^k \, dF(p)$.

It can be shown that this unique function is precisely that probability density function which we would have if we regarded the events as independent with a uniquely fixed objective probability. For finite sequences, a similar result holds, with summation rather than integration.

The immediate point of this result is to establish that any coherent assignment of probabilities to a sequence of exchangeable events can be modelled by a sequence of Bernoulli trials with uniquely specified prior probability. So, granted only coherence and exchangeability, our subjectively founded statistics will behave just as if we are dealing with independent events and objective probabilities, but without our being committed to some objective characterization of independence. We can

afford, like de Finetti (in one mood), to be agnostic about physical
chance: [1974: i. 218]:

> even if we were to consider someone's arguments to constitute an acceptable
> basis for an objective meaning of probability, our thesis consists in believing
> that these arguments would be irrelevant . . . because the logic of uncertainty is
> complete and clear as it is.

Exchangeability, then, provides a subjective replacement (an objectivist might call it a surrogate) for independence and symmetry in general. It is important to note that the representation theorem applies
without reference to frequencies. Suppose we are assigning probabilities, in advance of tossing a coin about which we have no previous
evidence but which we have decided to regard as fair, to its landing
heads on each of a hundred tosses. For the subjectivist, that simply
means that we have decided to treat the events as exchangeable and
representable by a Bernoulli sequence of degree 2. As a matter of fact
most of us are disposed to make such judgements; but objectivist theory
has great difficulty in making sense of decisions to treat a coin as fair
or unfair in the entire absence of frequency information. Is fairness a
physical property of the coin, expressible in terms of weight distribution, shape, etc.? Or is it a dispositional property which has never been
actualized? In making such judgements, are we making judgements
about causal links: then why identify probabilities with relative frequencies rather than propensities to display certain dispositions?

(*ii*)

Subjectivism does of course have an obligation to explain the role of
frequencies in our probability assessments; it does have a space for frequencies—that that space is narrower than in most objectivist theories
merely indicates that those theories 'restrict the treatment of uncertainty
to cases in which it can be presented in such a watered down way that
it looks like something else' (de Finetti [1974: i. 154]). Earlier subjectivists tended to treat our known reliance upon frequencies psychol-
ogistically: even in the famous 1937 de Finetti paper—which first proved
the representation theorem—we encounter airy references to 'profound
psychological reasons' why most of us base our estimates on frequency
information. One of the main achievements of de Finetti's later work has
been to replace this psychologism by an extension of the representation
theorem to demonstrate the rational compulsion for a coherent agent to
rely upon frequencies—a logical, not a psychological, compulsion.

Suppose you have an exchangeable set of events A_1, A_2, ... over which you have a subjective probability distribution P, equivalent, by the representation theorem, to a function $F(p)$ for a unique parameter p. Now, if you observe the outcomes of trials T_1, T_2, ... T_n of the n events A_1 ... A_n, you can partition the set $\{T_1 ... T_n\}$ into two sets H, containing those trials with outcome A_i, and K, containing those trials with outcome not A_i. Suppose the number of trials in H is h and in K, k. Then if we consider the unobserved events A the distribution of your probability evaluations conditional on your knowledge of H and K will, if you are coherent, remain exchangeable; and it can be shown (de Finetti [1974: ii. 221]) that these probabilities are the same as would be generated by $F(p)$, yielding the result: $F(p)$ is proportional to $p^h(1 - p)^k$. If we norm this distribution over [0, 1] by dividing through by n (equal to $h + k$), our posterior distribution for successes of the trials will have mode $h/h + k$—which is precisely the observed 'success frequency'. The larger the number of observations the closer to the mode our probability-values must be.

This is exactly the result at which a frequency theory would arrive by assuming the events independent and defining probability as a limit of relative frequency: the strong law of large numbers. But subjectivism has the use of this result without being constrained by it as to the objects of probability-appraisal. Assuming only exchangeability, wherever extensive observation yields a sharp estimate of relative frequency our prior probability-judgements, whatever (coherent) values they may have taken, will, when conditionalized upon these observations, produce a posterior distribution in agreement with that frequency. But unlike the frequentist the subjectivist can deal just as easily with cases of small or zero frequency information—since, as is clear from this result, frequency considerations are a means of assessing probability, not a criterion of its meaning.

It should be noted, though, that rather more is required for this second result than for the representation theorem. In contexts where exchangeability holds, coherence alone suffices to ensure that sequential judgements can be represented by judgements of objective independence. For the link with frequency, we need in addition the concept of conditional probability; the mechanism of conditionalization; and diachronic, as well as synchronic, coherence.

This additional theoretical apparatus becomes important when we ask whether exchangeability is a subjective or an objective matter and whether there is ever a compulsion upon a coherent rational agent to judge a set of events exchangeable.

(c) *Exchangeability: Subjective or Objective, Compulsory or Not?*

The ability to incorporate symmetry and frequency considerations, in their proper place, is a development of subjectivism which, if it succeeds, greatly increases the theory's appeal as a complete univocal interpretation of probability. But there is an ambiguity in the definition we have of exchangeability; and this ambiguity, once exposed, threatens to turn into an unpleasant dilemma for strong subjectivism. As we have seen, a set of events is exchangeable iff the measure of rationally warranted partial belief that a number of them will occur depends only on the number, not on which events make up the number. Is exchangeability then subjective, a property of our beliefs, a matter of order failing to affect our probability-judgements? Or is it an objective property of the set of events, which compels our judgements so that we cannot rationally take the set to be non-exchangeable?

De Finetti vacillates between these alternatives. He claims [1972] that no one is ever compelled to judge any set of events exchangeable: exchangeability is a property of the way our opinions are distributed, no more. But he later insists [1974: ii] that exchangeability is an objective feature of a given situation, which may often be determined with certainty, not a part of the 'logic of prevision'. Such objectivity, he claims, does not imply that probabilities themselves are objective, since they have, in exchangeable contexts or not, the same ontological status as any other opinions.

This latter position is not a tenable one for the subjectivist. Consider an example. Prior to a chess match of, say, ten games between two players, let us say Short and Fischer, we wish to make judgements of the probability that Short will win each of games one to ten. On an objective construal of exchangeability, the events are exchangeable if, as a matter of fact, order is unimportant; we can determine this fact with certainty (let us suppose) by considering factors such as age, stamina, resilience after loss, etc. for each player. Suppose we have determined the net effect of these factors to be nil, so that order does not matter (because, say, Fischer has less stamina but is more stubborn, and so on). Then the events are exchangeable, and, from the representation theorem, it would be irrational to hold a different subjective probability that Short will win the first three games from that that he will win the last three games. Now this conclusion is disastrous for strong subjectivism: a constraint other than coherence has to be admitted for monadic judgements, whereby something in the world renders some coherent distributions mistaken. A personalist such as Mellor could tolerate such a result, taking it to show that, in some cases at least, probabilities are the degrees of belief made reasonable by objective dispositions. But that simply means the abandoning of the central subjectivist thesis.

We are left with the alternative, subjective, treatment of exchangeability as a property of our beliefs. Suppose that a magician waves his wand over the chess-players to make Fischer as young as Short, Short as determined as Fischer etc. In every respect which we can even imagine as possibly relevant to the issue of who wins when, the players are indistinguishable. Still, since the magician has not tampered with our beliefs we remain free to treat the events as non-exchangeable and adopt any coherent distribution, assigning values of 0.9, say, to Short's winning games 1–3, only 0.1 to his winning games 8–10. Prima facie, this is an implausible account of probabilities. Moreover, it deprives the subjectivist account of consensus of its purported logical force. Being Bernoulli-modellable, or compatible with frequency information, is a property of your judgements which will obtain if you have allowed exchangeability to constrain those judgements. But you are never compelled to admit exchangeability. So, it would seem, divergent coherent judgements may rationally remain apart, however strong the accrued evidence. Serious as this difficulty is, this latter option is the only one possible if subjectivism is to remain subjective.

2.3. CONVERGENCE AND CONSENSUS

(a) *A Standard Subjectivist Account*

(i) Provided coherence is a sufficient condition for rationality and provided that coherence constrains conditional probabilities then Bayesian conditionalization and exchangeability will ensure that as evidence accrues distinct distributions will converge. This will be true in general whatever the nature of the events being judged probable. If, moreover, there is evidence as to relative frequencies then the distributions will converge towards the observed relative frequency and their variance from the frequency results will decrease as the number of observations constituting the background evidence increases.

(ii) We can now, in the light of the discussion so far, disentangle the elements in this picture which are clearly logical consequences of coherence from those which arguably are not. §2.1 has shown how if you are coherent you can preserve coherence only by changing your beliefs in accordance with a (generalized) rule of conditionalization. §2.2 has given reasons for supposing that coherence alone is never sufficient to compel you to judge any set of events as exchangeable. Hence, it seems that this account of the mechanism of convergence of opinions explains why there must be a consensus among rational agents who agree on judgements of exchangeability, but is not a full explanation of such consensus: for that we need in addition an account of the route from coherence to shared judgements of exchangeability. Furthermore, as §2.1 indicated, any such account must also encompass explanation of what constitutes evidential grounds for a change of belief in some way other than through objectively justified acceptance rules.

As has been made clear, the really serious problems for subjectivism are the problems arising out of (ii); before turning to them, let us look briefly at some objections which have been raised against the general programme of (i).

(b) *Three objections*

(*i*)

Horwich [1982] points out that it is always possible to argue backward, via Bayes's theorem, to produce coherent distributions so extreme that the effect of the accrued evidence will be inadequate to bring them in line with less extreme prior distributions, once we know how extensive that evidence is. The subjectivist account is inadequate, he claims, because it allows such extreme distributions to be rational even though they stand outside the consensus.

But this is to misunderstand the point of the account. First, it is a strength rather than a weakness that extreme distributions are not ruled out as irrational. The point is to explain why, in many cases, consensus is attained, not to argue that it must be; any theory which did the latter would be interpreting rationality implausibly strongly. Secondly, even these extreme distributions will still converge towards other distributions as evidence is taken into account; subjectivism cannot be taken to claim that for any evidence and any prior distributions the posterior distributions will ultimately agree.

(*ii*)

Objectivists such as Mellor argue that there is an onus on subjectivists to show how shared perception of chances could come about if there were no objective ground for agreement. This demand presupposes that the objective–subjective distinction should be interpreted ontologically, or at least externalistically—else what is a 'ground' of agreement? But if what is 'subjective' in subjectivism is an epistemological commitment then the burden on it is rather to demonstrate intersubjective belief sharing. It could be criticized if it failed in that task, but scarcely for failing to achieve what it is committed to avoid attempting.

(*iii*)

Inverse conditionalization will not, in general, enable us to choose between competing hypotheses both of which entail, or probabilify to the same extent, the observed evidence—in these cases the application of Bayes's theorem yields posterior probabilities for the hypotheses identical with, or in the same ratio as, their prior probabilities. Hence it is claimed to be inadequate to account for our preferences between theories in the large class of cases which involve choice between incompatible hypotheses consistent with and underdetermined by the data.

This objection does have force. It does not undermine conditionalization as an explanation of how consensus is arrived at; but it does suggest that the short Bayesian way with scientific method—equating it with conditionalization—is too short. I develop this point in §4.2.

(c) *Refining the Subjectivist Account*

Let me begin by making explicit an aspect of coherence implicit in much of the earlier discussion: that this notion presupposes a distinction between rationality and vindication of partial beliefs (cf. van Fraassen [1980*a*, 1983, 1985]). Suppose, invited to bet on a series of coin-tosses, you adopt a prior probability for heads on each toss of $p = 0.999 \ldots$ to 10^{100} places, with $1 - p$ for tails. You also agree that the series of tosses is exchangeable. You conditionalize your judgements correctly in accordance with Bayes's theorem as evidence accumulates that the relative frequency of heads is tending towards 0.5. In every respect, you have complied with subjectivist rationality constraints. Almost certainly, over the first million tosses you will lose drastically: yet, as Weatherford [1982: 229] puts it, 'if years of this losing enterprise laid waste [a bettor's] large fortune and left his family destitute then we still could not accuse him of acting irrationally'. Coherence provides no guarantee that your judgements will be vindicated, only a guarantee that they are not guaranteed not to be vindicated (by your being open to a Dutch book). It is, as van Fraassen points out, rather a constraint upon unbridled irrationality than a limit of rationality.

But what, then, of the calibration theorem? Does that not show that any rational agent must assign probability 1 to his or her being fully vindicated? Van Fraassen [1984] claims that it does and that the paradoxes thus generated demand that conditionalization be tempered by recognition that all judgements are instances of voluntary engagement—belief is a matter of the will. I believe that such a drastic voluntarism can be avoided: for the moment, let me defer attempting to show how (see §2.4 and Ch. 3). However, van Fraassen has a second line of attack, articulated in his comments on Shafer's [1983] article. If we combine calibration with conditionalization according to Skyrms's Rule C, then such conditionalization (upon evidence taken to be certain) forces the agent to believe that it is impossible that (s)he could take as evidence (bestow full belief on) a false proposition. It might be possible to extend the argument of the previous paragraph to meet this attack also; but, in any case, we can nullify the attack by discarding Rule C in favour of Jeffrey's Rule, as I suggested (on independent grounds) earlier. We need never then treat any evidence as certain.

With these refinements, we can preserve a thoroughgoing subjectivist account of belief change. Coherence, which guarantees rationality (but not vindication) of our judgements, is preserved iff we conditionalize upon evidence by assigning to it a probability as close as we like to 1, but never equal to 1. A further refinement stems from Teller [1973]; diachronic coherence requires conditionalization if we know in advance that we will acquire the information or its denial—otherwise we leave our conditional probabilities conditional. Dawid and Dickey [1977] cite examples suggesting that without this assumption conditionalization ceases to be compulsory. So, we can add to the subjectivist account a requirement that nothing is to count as evidence unless it has appeared as a term in our prior conditional distribution.

Subjectivism, then, is committed to opposing what has been called the 'revelation' model of evidence. Instead, it offers a picture of rational belief-change as a matter of an alteration in the monadic probability of some event E, upon which $p(E|E')$ had previously been assigned, consequent upon raising the monadic probability $p(E')$ from its prior level. As van Fraassen [1985] argues, it opposes not only the thesis that evidence exerts total control over opinion but also the thesis that evidence exerts total control over evidential judgement—that, if two agents have always taken the same propositions as evidence then they must have exactly the same judgements of evidence. Once that thesis is abandoned, it is clear that whether a judgement of evidence is shared by two agents is crucial to whether their distributions will converge. This is what lies behind Aumann's theorem [1976] that convergence depends on the existence of what Aumann vaguely called 'common knowledge': Shafer [1983] makes this notion precise by an axiomatization of the subjective concept of shared evidence.

The foregoing refinements to the convergence account and reflections on the nature of evidence point the way towards a solution of the deeper problem for exchangeability with which §2.2 ended and this section began. Take two coherent agents with differing monadic and dyadic probability distributions over a given probability space. By coherence, both are assured of rationality though they could not both be fully vindicated. As a consequence of coherence, the only rational way in which they can change their beliefs is by conditionalization, with evidence input as uncertain. If they now agree on an item of evidence— that is, assign the same monadic probability to the event that it has occurred—their distributions must converge. Exchangeability represents an agreement to assign the same monadic probability to each event in a set. If both agents do so, their distributions will converge onto relative frequencies. There is no need nor room in this account for objective acceptance-rules. But nor is exchangeability merely a matter of individual psychology: it is rather an intersubjective property of certain situations where there is agreement on what to count as evidence. Coherence leads inevitably to convergence and ultimate consensus. Granted this intersubjective property, that consensus logically cannot be other than in accord with relative frequency, where frequency information is available. And that result obtains while our account of probability remains wholly and strongly subjectivist.

2.4. WEIGHT OF EVIDENCE, RESILIENCY, AND SECOND-ORDER PROBABILITIES

This strong coherentist subjectivist theory (SCS) shows how coherence alone leads to ultimate consensus in our probability-ascriptions. But is such an enterprise, carried through with whatever degree of mathematical sophistication, sufficient to characterize uncertainty? Consider an analogy with ethical attitudes. You and I agree, let us suppose, that the issue of abortion on demand is a moral issue. You have brooded long and hard over the question, searching your conscience, comparing possible attitudes here with your attitudes on other moral issues; but, in the end, your attitude is just about as far from enthusiastic commendation as from outright condemnation. I agree that there is an important moral issue here, but I just have not bothered to form an attitude towards it. If our attitudes could be represented on some scale from commitment to abortion on demand's being a Very Good Thing to commitment to its being a Very Bad Thing, our positions might be assigned the same place near the middle. Yet no one would want to say that there is no important difference between you and me. They, most of them, will insist on commending you and disapproving of me. Similarly, it would seem that there may be important differences between two belief-states, each of which can be represented as degree of belief $\frac{1}{2}$ in some proposition, where one is the result of blank ignorance while the other is the result of possessing equal grounds for belief in the proposition as in its negation. We want to be able to claim that a preference for better-evidenced belief-states can be rationally supported, indeed is a factor in rational decision. Let us recall that Runyon's character The Sky lost most of his money as a young man through betting that St Louis was the largest city on earth. Given the evidence of his experience such a bet was not unreasonable; but we feel that a rational gambler ought to have weighed up how extensive his experience had been before taking on such a bet.

So the problem from which I start is that considerations such as these suggest the need for a parameter other than probability or rational degree of belief as part of a proper characterization of states of uncertainty.

This is of course an old problem. (*a*) presents it in the form in which it appears in Popper's 'paradox of ideal evidence', in Keynes's 'weight of evidence', in Jonathan Cohen's Baconian development of Keynes's notion, and in the 'epistemic reliability' idea advocated by Gärdenfors and Sahlin.

In (*b*) I outline the consequences of admitting such a parameter—which are dangerous for any univocal interpretation of probability, but potentially deadly for a strongly subjectivist position. In particular, I outline the force of the concession such an admission is to one part of Cohen's elaborate network of pluralisms—that concerning the role of probability in inductive inference and in judicial decision-making.

In (*c*) I change the subject: I return to the seemingly pleasant position at which the SCS programme had arrived at the end of the last section. But here, too, a problem emerges. To carry this interpretation through requires reliance on the notion that resiliency (roughly, stability) of our judgements is a factor constraining the rationality of those judgements— one which, contrary to the theory, does not seem to emerge out of coherence.

In (*d*) I examine ways in which subjectivism can incorporate the notion of second-order probability, subject to a generalization of a co-herence constraint.

(*e*) establishes a connection between second-order probabilities and resiliency of first-order distributions.

Then it turns out in (*f*) that these connected ideas make sense of weight of evidence while defusing its threat to subjectivism.

(a) *The Problem*

Recognition that there is a problem here goes back as least as far as Peirce:

to express the proper state of belief, not *one* number but *two* are requisite, the first depending on the inferred probability, the second on the amount of know-ledge on which that probability is based. [1932: ii. 421]

Keynes, too, in the course of developing his logical interpretation of probability, noted the existence of the difficulty, though, since he could not see any way to develop the concept of weight, he felt safe in declaring it to be of no practical consequence for action and so no threat to his account of the relation between evidence and the probabil-ity of an argument. He writes [1921: ch. 6]:

One argument has more *weight* than another if it is based upon a greater amount of relevant evidence; . . . It has a greater *probability* than another if the balance in its favour, of what evidence there is, is greater than the balance in favour of the argument with which we compare it; . . . As the relevant evidence at our disposal increases, the magnitude of the probability of the argument may either decrease or increase, according as the new knowledge strengthens the unfavourable or favourable evidence; but *something* seems to have increased in either case,—we have a more substantial basis upon which to rest our conclu-sion. I express this by saying that an accession of new evidence increases the *weight* of an argument. New evidence will sometimes decrease the probability of an argument, but it will always increase its 'weight'.

Popper, in an appendix to *The Logic of Scientific Discovery* written in 1958, puts the point by means of a forceful example (which derives from Peirce):

Let z be a certain penny, and let a be the statement 'the nth (as yet unobserved) toss of z will yield heads'. Within the subjective theory [which Popper takes to include logical theories], it may be assumed that the absolute (or prior) probability of the statement a is equal to $\frac{1}{2}$ . . . Now let e be some *statistical evidence . . . ideally favourable* to the hypothesis that z is strictly symmetrical . . . We then have no other option concerning P(a, e) than to assume that P(a, e) = $\frac{1}{2}$. This means that the probability of tossing heads remains unchanged in the light of the evidence e; . . . Now this is a little startling; for it means, more explicitly, that our so-called '*degree of rational belief*' *in the hypothesis, a, ought to be completely unaffected by the accumulated evidential knowledge, e*; that the absence of any statistical evidence concerning z justifies precisely the same 'degree of rational belief' as the weighty evidence of millions of observations which, *prima facie*, support or confirm or strengthen our belief.

This he claims, is paradoxical—the 'paradox of ideal evidence'.

In recent years, L. Jonathan Cohen [1977, 1985, 1986*a*, 1989] has taken up Keynes's notion in order to draw supporting parallels with his own theory of inductive support and inductive probability. As in Keynes, the weight of an argument and its probability are independent of one another. For Cohen, probability—of whatever kind—is a generalization of deductive inferability, the familiar mathematical 'Pascalian' probability being the result of one set of ways in which to generalize. The weight of an argument is 'the degree of derivability of an implicitly singular monadic probability judgment, such as the derivability of $p(Hs) = n$ from premisses $p(Hx|Ex) = n$ and Es', p here being Pascalian probability. Then '$p(Hx|Ex) = n$ is necessarily equivalent to $p(\text{not-}Hx|Ex) = 1 - n$ and $p(Hs) = n$ is necessarily equivalent to $p(\text{not-}Hs) = 1 - n$. From this it follows that the derivability of $p(\text{not-}Hs) = 1 - n$ from premiss $p(\text{not-}Hx|Ex) = 1 - n$ and Es must be just the same as the derivability of $p(Hs) = n$ from premisses p $(Hx|Ex) = n$ and Es, even when both derivabilities are very low, or very high, and therefore not mutually complementary. Correspondingly the weights of the arguments from Es to Hs and from Es to not-Hs must be reckoned the same. Hence the additivity of Pascalian probability in effect ensures the non-additivity of weight. Weight is therefore a non-Pascalian parameter of evidential strength' [1989: 103–4].

Recently, too, what I might call the Lund school (Gärdenfors, Sahlin, and others) have advanced models of decision situations as containing two components: an agent's set of subjective probability distributions and the epistemic reliabilities for that agent of those distributions. The simplest cases will be those where only two outcomes of each event are under consideration, so that the probability distributions can be described simply by the probability of one of the outcomes for each event. As an illustration [1982], they ask us to consider an agent whom they call 'Miss Julie' (the Strindberg reference presumably being a Swedish joke, though I admit I cannot see its point), who is asked to bet on the outcomes of three different tennis matches. For each match, her partial beliefs are to be represented, not by a single subjective probability but by the set of all the measures over possible outcomes which are epistemically possible for her (i.e. are internally coherent and do not contradict items which she knows). In match A she is extremely well informed about both players and predicts it will be a very even match with some minor factor determining the winner. In match B she knows nothing at all about either contestant. In match C the only thing she knows is that one of the contestants is an excellent player, the other a novice, but she has no idea which is which. If her beliefs were to be represented by a single probability-measure—so that she had to choose one set of odds at which to bet on each match—then she would have to choose odds of 50 : 50 in each case. If they are represented instead by a set of epistemically possible measures, in each case it is clear that the mean of those measures will be 0.5. But, Gärdenfors and Sahlin argue, there is an element in each situation not captured by any such set of probability-measures—the epistemic reliability of each measure. So, they say,

Miss Julie ascribes a much greater epistemic reliability to the probability distribution where each player has an equal chance of winning in match A where she knows a lot about the players than in match B where she knows nothing relevant about the players. In match C, where she knows that one player is superior to the other, but not which, the epistemically most reliable distributions are those where one player is certain to win. [1982: 368]

Epistemic reliability here is clearly the same parameter as we have been calling weight, or very closely related to it; at any rate, it differs from the probability-measure.

(b) *The Threat from the Concept of Weight*

One could multiply such examples of arguments for the need to employ a decision parameter other than probability and utility (perhaps the Levi–Shackle measure of potential surprise, or Carnap's requirement of total evidence, for example). But let us enquire instead into the consequences of admitting any such concept into an account of partial belief.

Keynes, as we have seen, thought weight did not matter. Popper, on the other hand, was convinced that, and of how, it did:

The *fundamental postulate of the subjective theory* is the postulate that degrees of the rationality of beliefs in the light of evidence exhibit a *linear order*. . . . all attempts to solve the problem of the weight of evidence within the framework of the subjective theory proceed by introducing, in addition to probability, *another measure of the rationality of belief in the light of evidence*. Whether this measure is called 'another dimension of probability', or 'degree of reliability in the light of the evidence', or 'weight of evidence' is quite irrelevant. What is relevant is the (implicit) admission that it is not possible to attribute linear order to degrees of the rationality of beliefs in the light of the evidence: that there may be *more than one way in which evidence may affect the rationality of a belief*. This admission is enough to overthrow the fundamental postulate on which the subjective theory is based. [1959: 407]

The force of Popper's point may be seen by considering Jeffrey's attempted resolution of the paradox [1983: 196]. Jeffrey claims that, in focusing attention on the probability to be assigned to one toss, the puzzle misdirects attention away from the locus of difference between our prior and posterior belief-states: that the latter, given ideal evidence, must assign degree $(\frac{1}{2})^n$ to any proposition that all of n tosses will yield heads ($n > 1$), but the former may assign a different value to such conjunctions—indeed, it must do so if we are to hope to learn from experience. 'Any decision about the rate at which you hope to learn from experience', he goes on, 'corresponds to some initial set of estimates about where the objective probability is likely to lie'. A committed subjectivist can achieve the same results, he argues, if for 'objective probability' is substituted 'the subjective probability measure that in fact we should all come to in the end'. But this reply will not do. The reliance on our hoping to learn from experience is question-begging: the point of the paradox is that a single-parameter theory cannot account for such hopes, since it takes the effect of evidence upon us to be exhausted by its effects on the measure of that parameter. Calling an additional parameter, instead of 'weight', 'objective probability' or 'the subjective probability we should all in fact come to in the end' is no help either (if that *in fact* implies objective convergence): again, the point is that respect for evidence is being accounted for in some way independent of the ground-level degrees attaching to our beliefs and consequently a constraint on rationality of probability-judgements must be admitted which does not emerge from constraints on the internal consistency of those beliefs.

It may be possible to evade the problem in a manner suggested by
the Lund school, treating desired level of epistemic reliability as a
measure of the risk preference of an agent. This measure will not then
function as an additional probability-measure; but it will become necessary to define utility partially with reference to epistemic reliability. In
effect, the problem is cleared from the realm of judgement by being
shunted into that of decision. But this is a very unhappy shift. First,
however weight or reliability is to be explained, it is undoubtedly an
epistemic concept, and attempts to account for it or measure it in terms
of non-epistemic criteria must be suspect: risk preferences may measure an agent's motives, but surely the import of evidence is primarily
a matter of reasons for belief, not motives for action. Secondly, there
are strategic grounds for wanting to avoid such a move: one of the main
merits of the de Finettian subjectivism which I support is that, by
contrast with preference theories, the complete explication of probability which it promises allows us safely to leave utility as an undefined
primitive term until probability has been defined. If we admit that
considerations about weight impose some structure on utility much of
that advantage is dissipated.

The most damaging consequences of all, though, emerge from
Jonathan Cohen's position. In bald outline (which does no justice at all
to the complexity and persuasiveness of Cohen's theory) the argument
runs as follows. Cohen begins by finding what he calls 'monocriterial'
accounts of probability (e.g. a pure frequency theory, or strong subjectivism) defective rather than wrong—applicable to some types of judgement only; equally, though, casual pluralist tolerance which fails to
explain why such diversity emerges from the applications of one concept rather than many is an unsatisfactory dodging of the main task of
theorizing, in philosophy as it would be in science. He finds the unity
of the concept in its resemblance to provability: probability is a measure of partial inferability, an indicator of the degree to which inferences
of some sorts are licensed. Then, to quote Cohen [1989: 112]:

Probability-judgments may be either necessary or contingent. Probability may
be viewed as a function either of terms or of propositions. Statements of probability may be either extensional or non-extensional. They may be either general
or implicitly singular, and either counterfactualisable or non-counterfactualisable.
Differing analyses of probability are made possible by the variety of possible
combinations of such options that they can exploit. So all this pluralism may
be viewed as operating within the framework that is constituted by the unifying
conception of probability as a gradation of inferability.

We have here a pluralism about probability interpretation which is, so to speak, internal to the standard 'Pascalian' calculus. But Cohen's position involves a more radical external pluralism too. The partial provability framework generates non-Pascalian probabilities too, if the generalization is from incomplete rather than complete deductive systems (a complete system, in this sense, being one where any sentence of the system is inferable from the system's axioms just in case its negation is not inferable): for such 'Baconian' or 'inductive' probabilities will not conform to a complementational principle for negation or a multiplication rule for conjunction. Weight of evidence, Cohen argues, has just such a Baconian structure. Its importance lies in the contribution it makes to demonstrating that Baconian systems are not just a theoretical possibility but rather the most natural way to understand important patterns of non-deductive reasoning. Cohen concentrates on two such areas.

First, he develops a theory of inductive support which takes it to be a gradation of the legisimilitude (closeness to being a natural law) of universally quantified conditionals, parallel to the gradation of their substitution-instances by inductive probability—which is, or at least closely resembles, weight of evidence. So weight here is crucial to the claim that neither inductive support nor inductive probability can be reduced to functions of mathematical, Pascalian, probabilities.

Secondly, in one important and reasonably rigorous framework, that of judicial decision, the principle of conformity to the standard calculus will not fit the mode of reasoning which juries are enjoined to pursue, which instead conforms to a Baconian structure. While in some inductive contexts it may be the case that 'weight-orientated modes of reasoning are not intrinsically in any kind of conflict with probabilistic ones but can serve to complement them' [1986c: 277], the verdicts of triers of fact are, and ought to be, supported by Baconian assessments of evidential weight which render Pascalian probabilities of ultimate issues otiose.

I apply the analysis to be developed in this section to these areas in Chapter 4.

(c) *How Far will Subjectivism Take Us?*

Let us think instead for a while about the successes of the SCS univocal approach to probability. This has furnished us with a wholly subjective explanation of widespread intersubjective consensus about a large range of probabilities, defusing the worry that the coherence constraint is implausibly tolerant. We know that any coherent assignment of probabilities to a sequence of exchangeable events can be modelled by a sequence of Bernoulli trials with fixed probability—what an objectivist would call independent events. And we know that wherever a posterior distribution across exchangeable events is arrived at by Bayesian conditionalization on the outcomes observed of trials of those events, coherence requires that the posterior distribution for successes of all trials will have as its mode the observed 'success frequency': the greater the number of observations, the closer to the mode our probability-values must be.

But why change one's beliefs at all? If coherence is the only constraint on the rationality of belief why not simply ignore the evidence and retain one's prior distribution? The analysis so far has offered no reason to prefer opinions conditionalized upon a great deal of evidence to those, no matter how extreme, which have not yet been much conditionalized. It offers no reason to prefer to judgements which are, we might say, on a rational track those which have gone further along that track.

A partial answer can be given in terms of resiliency. The onus is on the radical subjectivist to explain how our preference for more heavily conditionalized judgements will precipitate out of coherence, rather than representing a hard-core realist preference for 'truer' approximations to a unique unknown correct probability-value.

We can identify the preference more precisely as a preference for judgements with higher resiliency over larger domains and with smaller scope. Suppose we have conditionalized $p(H)$ over a domain of sentences $\{P_1, P_2, P_3, \ldots\}$. Resiliency is the complement of the maximum 'wiggle' of $p(H|P_i)$ about some value α: resiliency of $p(H)$'s being α over $\{P_1, P_2, \ldots\} = 1 - \max |\alpha - p(H|P_i)|$. If we consider the monadic probability $p_t(H)$ at time t as having derived via conditionalization over the P-domain from an initial monadic $P_0(H)$, high resiliency over the P-domain will represent a great degree of stability of $p(H)$ during that conditionalization.

But that is not quite what is represented by our practical preference. What interests us is whether we can expect our judgements to remain stable over possible further conditionalizations over some further set of sentences $\{Q_1, Q_2, \ldots\}$—call that the scope of $p(H)$. And that depends on the extent, to our judgement, to which the P-domain has incorporated the features causally relevant to H, leaving few such factors 'untested' in the scope. (The notion of resiliency may usefully be refined in at least two ways. First, we could stipulate that the domain include no self-contradictory sentences, inconsistent sets of sentences or sentences which are logical consequences of *H* or not-*H*: this would simplify the discussion of the 'Miss Julie' match C situation. Secondly, we could adjust the definition so that whatever the value of α the resiliency ranged over the whole of [0, 1]: this might make identification of an appropriate weight-function more straightforward. I have avoided cluttering up the main issue with these refinements.)

This analysis offers a characterization wholly in terms of subjective probabilities of our stability preferences. They do not threaten subjectivism; but do they threaten the coherentist element of SCS? Are we not admitting here a rationality constraint additional to that of coherence? What is left dangling here is the justification for that psychological preference: and what that comes down to is the question of the justification of our inductive habits, of our expectations of stability within our projections. And this, alas, is just the problem of weight of evidence all over again.

(d) *Second-order Probabilities*

Direct consideration of the concept of weight suggests that subjectivism is mistaken and incomplete; and our best efforts at developing subjectivism do not seem to yield a full explanation within the theory of the effects of evidence on our judgements. What can be done about this?

Let me raise what may seem to be a quite unconnected question: can SCS be expanded to incorporate second-order personal probabilities? Given that I can have beliefs about my beliefs, and those beliefs can come by degrees, can I have partial beliefs about my partial beliefs which, under some constraints, can be treated as probabilities? And could that constraint be no more than a generalization of coherence? As I have pointed out, subjectivists have traditionally fought shy of the notion of higher-order personal probability. The later de Finetti takes this line; but in his earliest work he was less hostile to the notion: see §28 of his 'Probabilismo' [1931], which hints at a treatment of 'subjective values' of probability statements in terms of the relative stability of probability-judgements. It is this hint which the next two sections follow up.

Such subjectivist anxiety stems, I think, from the natural worry that second-order personal probabilities must simply collapse onto first-order ones. There must, indeed, be a constraint upon second-order probabilities if paradoxical results are to be avoided: a generalization of coherence so that second-order dyadic probabilities will be coherent with first-order probabilities forming either term of the dyad. (Just because they are probabilities, of course, an agent's set of monadic and dyadic second-order partial beliefs must be internally coherent, as must his or her first-order beliefs.) Skyrms [1980a] demonstrates how this generalization can be accomplished. Suppose we have a language L_1 comprising non-probabilistic propositions $p, q, r, \ldots$ on which is defined a probability-measure p_1; extend L_1 to L_2 by adding propositions of the form $p_1(r) = n$ or $p_1(r)\varepsilon I$, I being a subinterval of $[0, 1]$, and defining a measure p_2 over the propositions of L_2. (I gloss over here the issue of whether p_2 should be defined over L_2 or only over the complement of L_1 in L_2. There is an oddity about attaching second-order probabilities to $p, q, r, \ldots$, but it is unimportant for present purposes: one can as well think of every L_1-proposition as having attached to it two numbers, a p_1-value and a p_2-value, as of its having a p_1-value attached to it to which a p_2-value is attached.) To avoid immediate paradox we must require that:

$$p_2\{r|p_1(r) = n\} = n, \text{ or, for interval probabilities, } p_2\{r|p_1(r)\varepsilon I\}\ \varepsilon I.$$

The dyadic second-order probabilities do indeed collapse onto first-order monadic ones. But this does not entail that second-order monadic probabilities will similarly collapse.

Applying Bayes's theorem to $p_2\{r|p_1(r) = n\}$ yields:

$$p_2\{r|p_1(r) = n\} = p_2\{p_1(r) = n|r\} \times p_2(r)/p_2\{p_1(r) = n\} = n.$$

Treating $p_2(r)$ as identical to $p_2\{p_1(r) = n\}$—as suggested above—this gives us a definition of the intuitively obscure $p_2\{p_1(r) = n|r\}$ as identical to $p_2\{r|p_1(r) = n\}$. The monadic second-order probability, which we are expressing indifferently as $p_2(r)$ or $p_2\{p_1(r) = n\}$, is not, however, determined by the first-order values, although any dyadic probabilities formed from it must satisfy the generalized coherence constraint. This derivation makes it easy to see why the objectivist takes second-order probability as trivial. If $n = 1$ or 0, the Bayes' expansion will go through. But if $0 < n < 1$, so that the LHS is positive, there will be a contradiction: the RHS will be zero, since, for the objectivist, $p_2\{p_1(r) = n|r\}$ must be zero if $n \neq 1$. But the subjectivist has no such problem: the 'given' of dyadic probability does not mean 'given knowledge of the true value', so there is no difficulty in constructing pay-off tables which allow $p_2\{p_1(r) = n|r\}$ to be n, whatever the value of n.

So second-order subjective probabilities are conceptually legitimate and non-trivial. But how can they be determined? Once an agent has accepted a book of bets which measures his or her first-order judgements, what further possibilities exist which would enable us to measure the second-order judgements?

One way to respond to this question would be to query or reject its operationalist presuppositions. Skyrms [1980b: 118] complains that one 'need not be so rigidly operationist as to assume that the *only* way that one can gain evidence for a degree of belief is by making a wager'. Second-order probabilities are a refinement of the intuitive notion of attitudes towards first-order odds, and function 'as theoretical parts of an imperfect but useful psychological model, rather than as concepts given a strict operational definition'. The modifications of de Finetti's naïve operationalism for which I argued earlier, in §1.2, make this now a viable and attractive option. Another, also quite defensible, response is that of Mellor [1980], who treats first-order beliefs as unconscious, inarticulate dispositions measurable in principle by bets while second-order beliefs are inner states of conscious assent determinable by introspection. But one can surely use the notion of coherence to do even better than either of these responses. The collapsing coherence constraint on dyadic second-order probabilities of the type considered above means that they can readily be determined by combining pay-off tables for appropriate first-order bets. The problem is how to get at the detached monadic judgements $p_2(r)$, i.e. $p_2 \{p_1(r) = n\}$ for some n, and at dyadic judgements such as $p_2\{p_1(r) = n|s\}$ where s contains no probability-assignments but may be evidence bearing on $p_1(r)$. (I shall return to this latter case in (f).) The former can be resolved at once by noting that an agent's monadic p_2 distribution must be internally coherent too, but that what corresponds here to the gain–loss outcomes of a simple bet is finding oneself in a bet at particular odds. Suppose an agent can choose to enter bets on a binary-outcome (r/not-r) single event at a finite range of odds $p_1(r) = \alpha_1, p_1(r) = \alpha_2, \ldots p_1(r) = \alpha_k$. Then $p_2\{p_1(r) = \alpha_i\}$ represents the odds that agent can choose, coherently with the other bets entered—and not necessarily uniquely determined—in a bet where the favourable outcome is that the agent can now bet on r at odds α_i, the unfavourable outcome being that no such bet is open. This is equivalent to the fraction of a unit stake which the agent can, coherently with other wagers, put up to buy entry to that p_1 first-order bet.

As an illustration, consider a restricted version of Miss Julie's decision situation. Making some obvious approximations, and assuming that in each case the only p_1-values open are 0, 0.5, and 1, one natural—

but not the only rational—way for her to distribute her p_2-values would be:

in match A, allocate her whole stake of 1 to $p_1 = 0.5$, nothing to 0 or 1;
in match B, she could allocate one-third of her stake each to the p_1-values 0, 0.5, and 1; or she could allocate her stake in some other coherent way;
in match C, she could allocate half of her stake to $p_1 = 0$, half to $p_1 = 1$, and nothing to $p_1 = 0.5$; but we would expect her to be unhappy about betting here at any odds, if it can be avoided.

Clearly, further analysis is needed here; (f) will reconsider this case.

(e) *The Link to Resiliency*

How might second-order probabilities connect with the resiliency of first-order conditionalized probabilities? We should expect there to be a link: intuitively, as evidence is conditionalized upon, it not only affects the odds but also concentrates our attitudes to particular odds. What we would expect is that for a given domain, the higher the resiliency of a first-order distribution the more 'concentrated' is any second-order distribution coherent with it. That is to say, the second-order distributions coherent with poorly resilient first-order distributions will be 'flat' (i.e. uniform) while those coherent with highly resilient ones will have high second-order probabilities attaching to a few points and almost zero probabilities at other values. Our practical preferences then represent a preference for more concentrated second-order distributions which will stably remain so over further conditionalization on any interesting scope of $p(H)$. That is psychologically very plausible. If we take something on which we have no decided views at all—say, the content of next year's birthday present, or on which we should have no views—the guilt of the defendant at the start of a case where we are jurors—our state can be described as one of no particular commitment to any ground-level probability, no strong attitude to any particular set of odds. Clearly, we do in general find a state of suspended judgement an uncomfortable one; we value the greater concentration deriving from greater stability because it gives us a commitment to stand by in reasoning or acting.

This is, of course, a very rough-and-ready account of the matter. It is important, I believe, not to become too besotted with the desire for numerical precision here: it would be tempting to try to present these notions in such a way that p_2-values represented Neyman-type levels of confidence in interval estimation of p_1-values, for example—but then our account would be applicable only to situations where sampling methods were appropriate. What can be done is to clarify our general intuitions and, in the process, delineate the limits of the cases where precise calculation is possible.

Suppose that we are evaluating a single L_1 proposition r and that the domain of resiliency of $p_1(r)$'s being α is a sequential feedback set $\{P_1, P_1 \wedge P_2, P_1 \wedge P_2 \wedge P_3 \ldots\}$ fixed and identical for each α in [0, 1]. Suppose that the uniform coherent p_2-distribution gives, for all α, a value of β for $p_2\{p_1(r) = \alpha\}$. Then I suggest that one plausible representation of the intuitions of the previous paragraphs is to take $p_2\{p_1(r) = \alpha\} = k \times \beta$ x res $p_1(r)$'s being α, over the fixed domain, for all α. The constant of proportionality k here is determined by the coherence constraint upon p_2 which requires that the sum of all the permissible p_2-values be 1.

In most decision situations this suggestion will not produce precise p_2-values. The resiliencies may not be well defined or may be indeterminate for some α's. It may be more appropriate to conditionalize over different domains for different α's. Conditionalization may be identifiable only as some vague constraint on the direction of change from prior to posterior distribution. Caveats such as these will apply to most everyday and forensic decisions. But, as I have indicated, such imprecision is to be welcomed here: in most situations first-order probabilities are similarly difficult to assess, and one ought to be suspicious of any theory which assumed a greater ease of assessment for second-order judgements.

The cases where greater precision is possible are those where the domain of resiliency comprises linear combinations of exchangeable first-order L_1 propositions. This is, of course, typically the situation which an objectivist would describe as one of repeated independent trials of an event. In such cases resiliencies about the possible α values approximate smoothly increasing or decreasing functions of the domain size. The convergence result presented earlier generalizes into a version of Savage's famous [1954] theorem, that as conditionalization is based on more and more evidence, for exchangeable events, the probability that the probability of the hypothesis' truth will exceed α approaches 1, for all $\alpha < 1$. Generalized, this becomes: increasing resiliency, given first-order exchangeability, tends to concentrate p_2 onto fewer p_1-values and the p_2 of these values approaches as close as coherence permits to

1. (Normally p_2 will concentrate onto a value near 1 attaching to one p_1-value.) Where the domain of resiliency includes feedback of information about the relative frequency of forecast success, this result will yield a solution to the problems of calibration of subjective probabilities, as I shall show in the next chapter.

(f) *Weight Revisited*

There is one blatant omission from the previous section's analysis of second-order probabilities: I have so far (deliberately, for the sake of presentation) ignored the point made earlier that our preferences for better evidenced judgements can only be explained by reference to the scope, as well as the domain, of resiliencies. Stability is not just a matter of how much conditionalization 'wiggle' has already taken place but of whether we expect any relatively sizeable further wiggle over the scope propositions. Once again, a precise evaluation of this factor will be possible only in the exchangeable cases; there, since order of conditionalization does not matter and wiggle approximates to a monotonically decreasing function of domain size, we can evaluate (via a law of large numbers) the precise effect of domain size on p_2-values. In general, though, the best we can do will be to estimate, based on our judgements about relevant causal or projectible factors, the weighting to attach to the domain : scope relationship—a matter of pragmatic judgement as to how easy or difficult it would be for our judgements to be much altered by new evidence.

Avoiding spurious attempts at precision, then, my full proposal for assigning second-order probabilities, with the previous constraints and an added factor w ($0 < w < 1$) to represent what I shall call the *weighting* an agent attaches to domain versus scope, is: $p_2\{p_1(r) = \alpha\} = k \times \beta \times$ res $p_1(r)$'s being $\alpha \times w$.

Let us look at how this might work out in the pared down Miss Julie example from (d). Here, r is the proposition s_1, 'the first player wins'; $p_1(r)$ can take only the values 0, 0.5, or 1. The uniform $p_2\{p_1(r) = \alpha\}$ distribution likewise contains only three points, so $\beta = 0.\dot{3}$. In match A, Miss Julie's well-informedness about the even matching of the players can be represented as very high resiliency ($\simeq 1$) of $p_1(r)$ about 0.5 when conditionalized over the sequential feedback set of the evidence available to her: likewise, res $p_1(r)$ about 0 and about 1 will each be close to 0. Fullness of information implies a value of w near to 1. Then

$$p_2\{p_1(r) = 0\} \simeq k \times 0.\dot{3} \times 0 \times 1 = 0;$$
$$p_2\{p_1(r) = 0.5\} \simeq k \times 0.\dot{3} \times 1 \times 1 = 1;$$
$$p_2\{p_1(r) = 1\} \simeq k \times 0.\dot{3} \times 0 \times 1 = 0.$$

(k must here be taken as 3 to ensure $p_2(r)$ coherence.)

In match B, her total ignorance can be represented by taking both res $p_1(r)$ and w as indeterminate for all α: consequently, she may rationally adopt any coherent p_2-values—the uniform distribution is most natural, but it is not rationally compelled; which is just what should be expected from a coherence constraint in the absence of conditionalization.

In match C, the domain of resiliency contains only the two propositions 'the first player is an expert' and 'the first player is a novice'. w will be rather low. Res $p_1(r)$ about 0.5 and about 0 and 1 will be indeterminate. (Because the sequential feedback set immediately contains a contradiction, the conditional probabilities are undefined: the only odds which can be chosen to generate a non-negative pay-off are 0/0. So the resiliency is indeterminate.) So, just as in B, any coherent distribution of p_2-values is open to her. Here, I am in conflict with the Lund analysis. On my view a piece of evidence which is worthless for conditionalization should not alter one's p_2-values from what they were in a state of complete ignorance. It can only seem that they should if p_1 is taken to be an objective probability. Taking it instead as rational degree of belief, the evidence available to Miss Julie here gives her no reason to change her attitudes to any bets offered; she should continue to have no more confidence in any one bet than any other, if that was her previous position.

This offers both a clarification and a refinement of our earlier intuitions.

So the analysis offered here accounts for weight of evidence not as a parameter independent of probability but not, either, as determined wholly by first-order probabilities: rather, it is a function of second-order probabilities or, which is the same, of the resiliency of first-order probabilities and domain-scope weighting. Thus, the weight of evidence for $p_1(r)$ being $\alpha = f[p_2\{p_1(r) = \alpha\}, \beta] = g[\text{res } p_1(r)\text{'s being } \alpha, w]$; one plausible candidate for f would be the variance of $p_2\{p_1(r) = \alpha\}$ about β. The choice of function is relatively unimportant: the issue is whether this analysis can deal with the arguments of Popper, the Lund school, and Cohen suggesting that weight could not be a function of probabilities.

Popper's 'paradox' disappears once we appreciate that the posterior second-order distribution for his coin differs from its prior, even though the first-order distributions are identical. It is perfectly correct to claim, as Popper does, that the introduction of p_2-values means that degrees of rational beliefs in the light of evidence can no longer all be linearly ordered. But that would be a problem for subjectivism only if coherence entailed linear order and the generalization of it to p_2 conditional on p_1 shows that it does not.

The Lund school offer some support, in fact, to my account, since they concede (Gärdenfors and Sahlin [1982]) that epistemic reliability can be interpreted as second-order probability. As I have argued, such a step is preferable to taking weight into risk preferences. The main difference between their analysis and this one is that they seem to wish, implausibly and unnecessarily, to treat epistemic reliability as objective rather than part of an agent's judgement.

Cohen's internal pluralism loses one of its chief motivations if weight can be accounted for within a univocally subjectivist theory: an important plank is removed from the platform supporting identifying probability with partial inferiability. In so far as that platform supports his external pluralism that, too, is weakened—but, as a later section will suggest, only in its claim that induction should be understood in terms of non-Pascalian inductive probabilities, not in its claim that induction involves factors other than Pascalian probability. The account given here of weight, if successful, undermines one argument for the central partial probability notion of Cohen's system; that notion is intimately bound up with the argument that weights must be non-additive—indeed, were it not for Cohen's independent arguments for the central tenet, his argument would move in a circle from probability as partial inferability to weight as non-additive to weight as inferability in incomplete systems to probability as partial inferability. I consider some of those other arguments later. My point is, however, that the subjectivist analysis of weight as a function of second-order probabilities given here prevents the concept of weight from offering separate support to the Cohen schema. Weight is not shown to be 'Baconian' probability; but this still leaves open the prospect that the structure of the concept, in some contexts at least, can be analysed in something like a Baconian way—for it is not simply a function of Pascalian first-order probabilities only.

Finally, to tidy up an issue left dangling from (d): occasionally we are interested in establishing the weight of evidence for $p_1(r)$ being α

given s, where s is a proposition neither in the domain nor scope of res $p_1(r)$ about α. There is no difficulty in carrying through a similar analysis for conditional resiliency (since no paradoxes loom here), thus:

the weight of evidence for $p_1(r)$ being α, given $s = f[p_2\{p_1(r|s) = \alpha, \beta] = g[\text{res } p_1 (r|s)\text{'s being } \alpha, w]$.

That we can thus account for the weight of evidence for both monadic and dyadic first-order probability-judgements supports the contention which I advanced earlier (in §1.4), that Mellor's [1971] argument against conditional betting need not undermine the subjectivist treatment of conditional probability. As has been seen in §2.1, 'conditional bets' can best be seen as compound non-conditional bets. A coherent agent choosing a single p_1 value for a simple non-conditional bet on some proposition q will normally be taken to be—and, if possessing a rationally proper respect for weight of evidence, will be—choosing such a value in the light of all the evidence available concerning q. Similarly, when such an agent now assigns a probability-value to $q|r$, (s)he will be taking into account the way both the balance of evidence in favour of, and the total weight of evidence for and against, q will be affected by the addition of r alone to the agent's existing corpus of evidence concerning q.

3

From Subjectivism to Projectivism

As an increasingly rigorous account of SCS has developed up to this point, the theory has moved further and further away from the crude sketch of it presented at the beginning—a crude sketch which is too frequently what opponents of subjectivism take it to be. Given the tendency among these critics to assume that any theory more sophisticated than that is *ipso facto* less subjectivist, it is time now to enquire more closely into the precise nature of the *subjectivism* of the developed theory. The Introduction argued, partly from the failure of the objectivist alternatives, that there is scope for a personalist interpretation of probability; and Chapters 1 and 2 presented the core of a theory which attempts to build such a personalist account on the single, minimal, constraint of coherence. But, apart from the objectivist competitors, there are of course within a personalist framework options other than the de Finettian approach I have been adopting—it is important to see how it relates to these options and to indicate, at least, why it is to be preferred to them. Several forms of pluralism in the interpretation of probability also need to be considered, before moving ahead with the positive project of locating our subjectivist commitments as commitments to a projective—better, a quasi-realist—approach to the problems of probability theory.

§3.1 considers:

    (*a*) subjectivism which identifies itself with a minimal personalism
    (*b*) psychologistic personalism
    (*c*) personalist logical-relation theories
    (*d*) personalist propensity theories
    (*e*) personalist indeterminate-probability theories.

§3.2 examines various pluralist options:

    (*a*) some relatively unsophisticated, though well-known, theories
    (*b*) Cohen's radical 'Baconian' pluralism
    (*c*) the pluralist urging, emerging from 'probability-kinematics' or

'objectification' approaches, that both the concepts of subjective credence and objective chance are needed in a proper analysis of probability.

Several of these options have been given sophisticated and complex expression in thoroughly worked-out systems. This chapter does not pretend to provide a comprehensive critique of these systems (which would demand a critical analysis longer than the positive part of this book): rather, it attempts the lesser task of placing the theory developed so far—and with some inevitable references forward to later chapters— in the context of these alternatives so as to provide a clearer view both of its strengths and of the task to be pursued later.

§3.3 undertakes the consequent, and more important, positive project:

- (*a*) looks at the common misidentification of subjectivism's prime commitment as ontological.
- (*b*) offers a first sketch of SCS as outright anti-realism and looks at some consequent epistemological issues;
- (*c*) suggests that SCS need not be troubled by some generally trouble-some worries about the links between recognition-transcendence, bivalence, and determinacy;
- (*d*) suggests that a quasi-realist direction for SCS is preferable to an instrumentalist one;
- (*e*) indicates some elements of realist-seeming probability practice which SCS must take on the burden of explaining;
- (*f*) explores how a projectivism which makes sense of calibration, the rationality–vindication relationship, resiliency, and higher-order degrees of belief can claim to effect a quasi-realist enterprise of founding apparently realist practice on a subjectivist basis.

3.1. MISGUIDED DIRECTIONS

(a) *Personalist Minimalism*

Mellor [1971] makes this useful, if seemingly obvious, distinction between subjectivist and personalist theories of probability: to be a personalist is to hold that the partial beliefs of a real or imagined individual, manifested in (rational) betting-behaviour, can form the basis of an interpretation of the formal probability calculus; subjectivism is the stronger doctrine that *only* individual degrees of belief can be construed as probability-assignments and hence that objective chance is, if not conceptually impossible, at any rate redundant. In Mellor's own system, chance statements are expressions of an individual's partial belief, but what makes such statements reasonable or unreasonable is the corresponding objective propensity in the world—so this account of probability is both personalist and anti-subjectivist.

I describe this distinction as 'seemingly' obvious because, although it is undeniably correct and important, it is a distinction many subjectivists fight shy of making. Kyburg and Smokler, for instance, in their editorial introduction to the classic text *Studies in Subjective Probability* [1964] characterize as subjectivist any theory built around coherent, but not fixed, degrees of belief. Even in the most detailed and complete account, de Finetti's [1974], evasion and equivocation abound. On the one hand are clear, indeed bold, assertions of strong subjectivism:

only subjective probabilities exist, i.e. the degree of belief in the occurrence of an event attributed at a given instant and with a given degree of information [i. 3]

even if we were to consider someone's arguments to constitute an acceptable basis for an objective meaning of probability . . . our thesis consists in believing that these arguments would be irrelevant anyway. [i. 218]

Yet elsewhere are claims that the theory is neutral as to the existence of objective chance; and throughout, the work constantly envisages its opponents as attacking subjectivism by means only of arguments against the viability of personalism.

Clearly, the strong thesis of subjectivism cannot be earned on the basis of success in the weaker task of demonstrating the possibility of personalism. That many subjectivists have been tempted here towards theft rather than honest toil reflects, I believe, the fact that for so long the theory served the needs of statisticians while being ignored by philosophers. De Finetti [1974: i. 94] poses as the plain mathematician: subjectivism is 'not bound up with any particular philosophical position nor incompatible with any such . . . in order to avoid becoming embroiled in philosophical controversy . . . such as Determinism versus Indeterminism or Realism versus Solipsism'. The hope is to sidestep the thornier issues of adequacy and completeness by gesturing to the interpreted mathematical core of the theory, which indeed demonstrates that modelling the formal calculus in terms of coherent partial belief will permit development of models of conditionalization and convergence onto frequencies. But of course such gesturing is not enough: both pluralism and Mellor-type theories also incorporate that core.

(b) *Psychologistic Personalism*

Opponents of subjectivism frequently argue or presume that it must be essentially psychologistic, in the post-Fregean pejorative use of that term. Thus, Weatherford [1982: 238–9]:

A psychologism is the mistaking of a belief or opinion for a fact or objective condition, or the other way round . . . In probability theory, it is psychologistic to confuse people's beliefs and behaviour with the objective grounds of those beliefs . . . The most basic criticism of subjectivistic probability is that it confuses feeling with fact.

And, certainly, earlier subjectivist theorists (indeed, not only those of personalist inclinations) from Huygens through the Bernoullis and Bayes to de Finetti and Ramsey in the 1920s, did little to escape the charge that they conflate, without argument, what goes on in the world and what in the head. Two main areas of confusion leave room for this criticism.

First, there is the issue discussed (and, I think, resolved) in §1.5 of whether the theory is to be construed as purely descriptive or as purely normative: both unattractive options. The analysis developed earlier shows how SCS can provide an idealized description of actual practice, be informative about the mechanisms whereby the actual divergent judgements of imperfectly rational agents do, in practice, converge and yet have regulative force.

Secondly, early theorists were often confused or evasive as to whether, or to what extent, probability is a measure of an individual's knowledge or state of ignorance—their motivation in that direction often, of course, including a wholehearted determinism. At the same time, however, they wanted to see their theories as directly informative about the world. The simplest way to reconcile these aims was just to blur the impropriety of using the term 'probability' to refer both to an internal state of uncertainty and an external state of affairs despite the lack of a theoretical justification for doing so. And that can fairly be called a psychologistic manœuvre.

Thus Bayes, in his famous [1763] essay, having defined probability as 'the ratio between the value at which the expectation depending upon the happening of the event ought to be computed, and the value of the thing expected upon its happening', then equivocates upon 'expect', sometimes taking it as wholly relative to an individual's mental state, sometimes as though expectations were externally fixed values. Later [1931], Ramsey was certainly aware of the danger of psychologism:

he denied that chances correspond to anyone's actual degrees of belief and asserted that some opinions about chances are much better than others. Nevertheless (*pace* Blackburn [1980*a*]) he did not escape the psychologistic trap: in order to tie behaviour to the laws of probability, he advanced 'a general psychological theory . . . that we act in the way we think most likely to realise the objects of our desires, so that a person's actions are completely determined by his desires and opinions' and 'a law of psychology that his behaviour is governed by what is called the mathematical expectation' [1931: 172]. This is quite inadequate as a solution of the problem of the gap between theory and actuality—indeed this 'law' is empirically wrong. (See e.g. Davidson, Suppes, and Siegel [1957] or John Cohen [1972].) In a similar vein, the early (pre-1950) de Finetti took the line that 'probability theory is not an attempt to describe actual behaviour; its subject is coherent behaviour, and the fact that people are only more or less coherent is inessential' [1937: 103]. Most of us, he continued, do generally accept the rules of coherence even if we frequently violate them (just as we accept the laws of arithmetic even though we may make frequent errors in long division)—there are 'rather profound psychological reasons which make the exact or approximate agreement that is observed between the opinions of different individuals very natural' [1937: 152]: but de Finetti nowhere reveals what the profound psychological reasons are.

In the previous two chapters I have been concerned to 'de-psychologize' an extended concept of coherence—operating as both static and dynamic constraint on monadic and dyadic partial beliefs—into a minimal rationality constraint which is none the less sufficient to ground any judgements that interpret the calculus. Whether or not SCS can be agnostic about questions of determinism it must deal with both of the issues I have been discussing here. The theory has gone a considerable distance towards resolving them. But it is at least arguable (and argued very strongly by Vickers [1988]) that, as long as SCS founds its concept of probability on acts of judgement rather than their contents, it will remain rooted in psychologism—and, if it takes that further step, it will turn away from personalism towards logical-relation theory. I shall tackle this dilemma in the conclusion.

Modern personalism has responded to these accusations of psychologism in one of two ways. Either it has conceded that, although probabilities can be taken to be degrees of belief, these beliefs must be constrained by some objective standards which determine that some beliefs are correct while others are not—leading to one of the versions of personalism described below; or, as in de Finetti, it has attempted to develop the coherence requirement as objective, but not external, grounds for our beliefs.

This latter approach, I have been arguing, ought to be preferred as less committal: but de Finetti's own [1974] presentation of it can scarcely seem acceptable. He begins from the very strong assertion that probabilities do not exist outside the mind, but goes on [1974: i. 217] to claim that the theory is neutral as to the existence of physical chance; the condition of coherence is said [1974: i. 85] to be objective, but at the same time a characteristic only of an individual's subjective evaluations. This confusion is quite unnecessary. In the light of the discussion so far, it is simpler to present subjectivism as identifying probabilities with the beliefs one can hold within the constraint of coherence; falling within that constraint or outside it is grounds for accepting some belief distributions and not others as probability distributions—but, in general, one should not seek grounds which will restrict us to one permissible distribution only. The coherence constraint is external to any particular set of judgements but ultimately derived from our judgemental practices.

(c) *Personalism and Logical-Relation Theories*

Coherence, as we have seen, is a constraint upon all possible combinations of our partial beliefs: to avoid *ad hoc* escape from Dutch books, we must be supposed compelled to bet on any collection of events an opponent might nominate. Given this, and the logical interdependence of many of our beliefs, we may expect that the coherence requirement will in practice be a powerful one. Very often, indeed, it may mean that only one probability-value may coherently be assigned to a new proposition, given our prior distribution. This effect might be thought to bring personalism very close to a logical-relation theory (cf. Benenson [1984]) where the probability of a proposition is, or at least is fixed by, its relation with the body of evidence relevant to that proposition. Carnap [1950] was sufficiently impressed by the similarities to list the most prominent subjectivist theorists as members of the same logical-relation school as Keynes and himself (though later [1971] he came both to appreciate the contrasts and to move significantly towards subjectivism in his own theory).

The distinction between the two stances is, however, clear. For a subjectivist, probabilities are relative to the state of information of a given individual at a given time; for a logical-relation theorist, they are relative to the totality of relevant evidence, whether the individual is aware of it or not, and hence uniquely determined, however individual estimates of them vary. So-called hybrid personalist–logical-relation theories are no more than logical-relation theories simple, and face just the same problems: defining relevance and detecting or characterizing the measure-function of the probability-relation.

(d) *Personalist propensity theories*

In Mellor's [1971] system, the making of a probability statement in-variably expresses the speaker's partial belief in an event's occurrence, and this partial belief is measurable by betting-behaviour. But, unlike strong subjectivists, he does not allow equal status to all coherent sets of partial beliefs; instead he claims that there are propensities in nature which make certain partial beliefs more reasonable than others. His argument that there must be such propensities falls into two stages: first, he claims that a study of usage shows that chance statements are normally intended to convey more than an individual's beliefs and that there is such widespread intersubjective agreement on some chance statements as to give an objective ground for their prima-facie plaus-ibility; secondly, and consequently, the onus falls on subjectivists to show how this shared perception of chances comes about if there is no objective ground for such agreement.

I make no objection to the first part of the argument. Neither would de Finetti; indeed he explicitly quotes instances which suggest a wide-spread tendency towards belief in an objective ground for our statements of probability (e.g. [1937: 118] where he asks 'Why are we obliged in the majority of problems to evaluate a probability according to the observation of a frequency?'). As to the second part, Chapter 2 has provided a detailed response to this challenge: first, our theory shows that for sequential conditionalizations of our probability-judgements to remain coherent it is necessary that we use Bayes's theorem to conditionalize on the acquired evidence, and so any non-extreme dis-tributions will tend to converge; secondly, and most vitally in scientific contexts, when this process of Bayesian conditionalization is carried out on exchangeable collections of events it has been shown that the distributions will not only converge, but converge onto observed rela-tive frequencies—perceptions of which are shared and objective.

One major flaw in Mellor's attack on subjectivism is that he simply ignores this concept of exchangeability and so misses the point that in most cases of interest to science the subjectivist must, and can, use a relative frequency to arrive at a probability-judgement. So, when he attacks [1971: 45–9] what he calls 'the subjective surrogates for chance' he is in fact attacking a very much weaker theory: one in which Bayesian conditionalization alone, rather than together with the consequences of exchangeability, is used as the 'causal explanation of how scientists . . . are brought by the piling up of shared evidence into the close agreement that is observed in their chance assignments'. As a result, most of his criticisms simply miss their target. 'No one', he says, 'supposes the committees who settle and revise values of radioactive constants to proceed by conditionalizing coherent betting quotients.' Certainly the theory advocated here does not, instead supposing that these committees proceed by commissioning experiments—that is, sets of exchangeable observations—leading to averaged results, which of course represent frequencies.

(e) *Personalist 'Indeterminate Probability' Theories*

These theories, advocated (in different versions) principally by Kyburg and Levi, trace their origin to work by Good and Kyburg in the early 1960s. They begin by noting the implausibility of assuming that all of an agent's partial beliefs should be comparable in intensity so as to allow us to form a single scale of degree of belief and assign a single real number as a measure of intensity of each belief. Might it not instead be the case that probabilities should be interpreted as intervals within which the agent's degree of belief falls, or that an agent's partial beliefs can at best be represented by a set of credal states—not a single state—relative to some field of propositions? It certainly appears as though the personalist coherence requirement, however tight in practice, will still determine only an interval or set, not a unique value. But then something more than the degree-of-belief conception seems to be needed to interpret the formal calculus. Kyburg [1974] moves in the direction of incorporating direct inference from knowledge of relative frequencies as a constraint additional to coherence; both he and Levi [1980] move towards logical-relation theory, arguing that the credal state of an agent, in addition to an internal constraint of coherence, must be determined by its relation to the whole corpus of knowledge available.

It seems to me that these theories stem from a mistaken view of the way in which coherence regulates our judgements. For a subjectivist, the claim that, relative to an agent A, the probability of an event E is *k* implies that if A is ideally rational (s)he *may* choose *k* as his/her intensity of degree of belief in E—*k* is, so to speak, within his/her coherent range—and that *k* is the value (s)he *has* chosen. This claim does not imply that an ideally rational A could not have chosen a different value of *k* within the range. In extreme cases, such as a single-event universe with no background evidence, any value of *k* between 0 and 1 could be chosen. That there may be more than one coherent distribution of judgements does not undermine the claim of the measures of belief-intensity to conform to the formal calculus. The subjectivist can construe the starting intuition of these theories, not as implying that probabilities are indeterminate, but as reflecting the uncontentious fact that any coherent subject's probabilities might have been different.

3.2. PLURALISMS

(a) *Some Earlier Versions*

Even from the brief account of the Introduction it should have been clear that none of the objectivist theories considered could cope adequately with the three key questions asked there. So, it is not possible to give a complete account of our use of 'probable' in terms solely of symmetry, of logical relationships, of relative frequencies, or of physical propensities. But this does not exhaust the objectivist options; for we have looked so far only at theories proffering one monistic interpretation for all uses of 'probable'. And, to quote L. J. Cohen [1977: 7]:

in recent decades there has been a substantial trend of opinion towards some kind of polycriterial account of probability. Rival monocriterial theories claim to refute one another. But a polycriterial account supposes this or that monocriterial analysis, if internally consistent, to be not wholly refuted by the diversity of the facts—merely restricted in its domain of application.

One group of issues, glossed over earlier, seems to lend support to a pluralist approach. It might be thought important to make a radical distinction between the probability of a proposition or hypothesis on the one hand and the probability of an event on the other: between epistemic or inductive probability and statistical or causal probability. On various different ways of viewing probability, it is natural to regard the probability of an event as supervenient upon events in the world—

which do not include possession of, or lack of, items of evidence. That is, the probability of an event cannot alter unless some event-like particular alters; but the probability of a proposition would seem to be affected just by the acquisition of evidence relevant to the proposition. Without some such distinction it is very difficult to make sense of tensed probability assertions. But one should note that even if there is a difference in supervenience conditions, that need not entail a difference in meaning. To hold that the sense in which a proposition may be probable differs from the sense in which an event may be probable renders problematical a very common pattern of reasoning, e.g. (from Ayers [1968]):

The hypothesis 'all red morning skies are followed by rain' is very probable.
This morning the sky was red.
So, the event of rain today is very probable.

Justified or not, this distinction has inclined several writers towards various kinds of pluralism. Perhaps one or several of the interpretations considered in the Introduction is appropriate to the different cases of probability of a proposition and some other interpretation(s) to cases of the probability of an event. A question of meaning then arises: are these different cases cases of different concepts or is there a unitary meaning with different applications?

I take Carnap's early position to be not far removed from the extreme of supposing that the different applications of the term 'probable' in usage are mere homonyms, with no common core of meaning at all. He claims [1950: 19] that

there are two fundamentally different concepts for which the term 'probability' is in general use ... probability$_1$ is the degree of confirmation of an hypothesis h with respect to an evidence statement e ... probability$_2$ is the relative frequency (in the long run).

Carnap accounts for the existence of one term with two fundamentally different meanings as a consequence of the development of a theoretical approach to probability refining and explicating a pre-scientific concept. To compare this account with von Mises's, discussed earlier, consider the term 'work', which has a pre-scientific meaning characterizable by its connotations of effort, labour, duty, etc. When mechanics extracts from this term the definition 'work = force × distance' it is refining one single concept from the pre-scientific quarry; but other instances, such as the notion of the *vis viva* of a body current in the early seventeenth century, turn out to be refinable into more than one concept—here, energy and also momentum. So it is, Carnap would argue, with the pre-scientific concept 'probable' (equivalent, perhaps, in Aristotle's phrase, to 'what for the most part happens') which yields his two distinct concepts.

But the trouble with such accounts is that they risk degenerating into mere etymology—and often rather dubious etymology, too. Any satisfactory account of probability must surely include, even if implicitly, an explanation of why the term 'probable' has such diversity of usage: an explanation which goes beyond the merely historical to argue the appropriateness of using the same word for distinct theoretical or prototheoretical purposes.

So is it possible to combine tolerance of pluralist usage with an underlying theoretical unification? The most sophisticated and radical attempt to do so has been made by L. J. Cohen; it raises so many important issues that I consider it separately in the next section. Three other 'unitary' pluralisms have been suggested which I shall consider briefly here: none holds out much hope of providing a genuinely explanatory theory.

First, a formalist account would attempt to depart as little as possible from what Black [1967] called the 'mathematical dogmatism' view: that no definition, other than a formal one, of 'probable' is necessary or possible and that probability theorists should confine themselves to utilizing the mathematical theory. The formalist account does not go quite so far, but would limit theories of probability to simply being statements of the calculus together with a list of criteria (betting, frequencies, etc.) showing why each function that conforms to the criterion must be a probability-function. But, as L. J. Cohen [1977] points out, to do this is not to solve any problems of interpretation but merely to evade them; it makes no progress towards showing why this particular piece of mathematical syntax should have corresponding to it a diversity of semantical theories, each having some claim to usefulness as an interpretation of the calculus and a guide to action.

Secondly, there is the account offered by Mackie [1973], according to which 'probable' has at least five basic senses, not accidental homonyms but rather possessing a family resemblance brought about as some common ancestor spread from context to context. But, again as Cohen points out, this account fails to be genuinely explanatory (unless it is construed as pure etymology, in which case it is just wrong). For some groups of . . . sorts have common names, and some do not, even when the sorts exhibit family resemblance to one another. So unless the exponent of a family-resemblance approach to probability tells us why *his* supposed nexus of family resemblance generates a common name, in contrast with others that do not, he has not explained anything. But if he does do this, and does it adequately, he has to go beyond a merely family-resemblance account. [1989: 83]

Even the most committed Wittgensteinian, happy to accept family-resemblance accounts of everyday common nouns and suspicious of the desire for explanation within philosophy, is unlikely to be so happy with the account of a term which has undergone the process, familiar in scientific and mathematical contexts, of abstraction from everyday practice into a formal system which is then 'read back' into the practice. (Wittgenstein himself may well have excepted this type of case from his family-resemblance commitment, at least if we can take Waismann [1965: 183] to represent his views.)

Thirdly, Good [1962] has suggested that the underlying unity is to be found in seeing all probabilities as relative to the state of information of some organism (except that there may or may not be physical probabilities). Apart from these, if they exist, there are three organism-relative types: psychological probabilities, that is, values directly observable in behaviour; subjective probability, that is, 'psychological probability modified by the attempt to achieve consistency'; and logical probability, that is, the hypothetical subjective probability of an infinitely rational being. The organisms referred to need not be human—they might be machines or Martians or information systems. Divested of its science fiction overtones, this idea seems to me to contain nothing which cannot be contained in a personalism which generalizes from ordinary practice to regulative principles which it then connects with conceptions of ideal rationality and ideal correctness. Unitary understanding is certainly within Good's grasp, but only via effectively abandoning pluralism.

(b) *Cohen's Radical Pluralism*

None of the pluralist accounts just considered could offer the explanatory power needed to present different interpretations as possessing an underlying conceptual unity. Unless we are prepared to rest content with Ayer's [1972: 27] quietism—'I take no side on the question whether there is one or more than one concept that goes under the name of "probability", because I am not sure what the criteria are for individuating concepts and suppose that they are rather arbitrary'—something more sophisticated is needed. Jonathan Cohen in several books [1970, 1977, 1986*a*, 1989] has developed a powerful systematic theory which leads him into pluralism at several levels of interpretation. I have already outlined this theory in the discussion of weight of evidence in §2.4; let me now consider it in rather more detail.

Cohen's general strategy may be divided into three parts. First, he claims that in any framework of measurement, including probability, one must distinguish the criterion of gradation used from its methods of assessment and ask whether we have a plurality of criteria or one criterion with a plurality of means of assessment. Secondly, he argues that to avoid the anomalies of a monocriterial account of probability, yet do justice to the fundamental unity of the concept, it is possible and natural to found our theory on a conception of probability as generalized, partial provability. Thirdly, he affirms the usefulness of the explanatory theory by showing that it predicts the possibility of situations involving criteria of probability which have a syntactic structure different from the standard calculus—and then claims that inductive reasoning and judicial decision are two such situations.

As regards the first of these, I am happy about distinguishing definition of a concept from mode of assessment (as may have been evident in my discussion of AP theories), but I find Cohen's characterization of 'criterion of gradation' rather unfortunate. He compares the interpretation of probability with the nature of temperature measurement. We distinguish a pattern of ordering such as the Celsius scale from the various methods of assessment (mercury thermometers, thermocouples, etc.) which allow us to assign to some item a location on the scale. A weak, generally harmless, form of pluralism would be the claim that some criterion of gradation possesses various means of assessment; much stronger and less uncontentious would be the claim that some concept properly supports, or is ambiguous between, several criteria of gradation, e.g. 'quantity' of apples might have reference to weight, volume, number . . . In the case of probability, Cohen suggests that the diversity of interpretations reflects the latter, stronger type of pluralism (though of course the weaker may also be in play). The unfortunate thing about this analogy is that the notions imported from measurement are themselves unclear because many more than two such notions need to be distinguished there—acts of measuring, results of measuring acts, scales of measurement, ordering relationships, quantity, magnitude— which the analogy compresses and confuses. Still, this is perhaps not a very serious fault in Cohen's theory: read 'polycriterial' as simply 'pluralist' and the argument will lose little if any force—though the location of the pluralism will then remain to be determined.

Consider now the second element in the Cohen strategy. Why should probability be regarded as a generalization of provability?

There is a result established by Popper [1959: app. *iv and *v] which appears to provide a link between probabilities—at any rate if treated as, in at least some contexts, relations between propositions—and deductive entailment. Popper takes conditional probability as basic; absolute probability is simply the probability of a proposition conditional on a tautology. He then defines legitimization thus: A legitimizes B means that, for all C, $p(A|C) \leq p(B|C)$; A is said to be equipollent to B if A legitimizes B and B legitimizes A. It can be shown that equipollence is a equivalence relation and that the equivalence classes of equipollent sentences form a Boolean algebra with the class of certain propositions as the maximal element and the class of impossible propositions as the minimal element. Popper concludes that 'since a Boolean algebra may be interpreted as a logic of derivation, the probability calculus is a genuine generalization of the logic of derivation'.

There are two ways of looking at the significance of this result. Either one can take it as showing that conditional probability-functions can provide a semantics for deductive logic—as Field [1977] does, though this is not germane to the current issue—or one can take it— as Popper does—as showing that probability is a relation analogous to entailment, hence justifying regarding probability as degree of inferential soundness.

This latter interpretation tries to squeeze too much from the theorem. For, first, Popper's analysis depends upon assuming the probabilities in question to be logical functions whose arguments are sentences; so at most he has shown that so long as probabilities are interpreted in that manner their equivalence has a structure similar to deductive entailment. But that is of little service to pluralism. Secondly, even that contention is being stretched if it is taken to mean that probability generalizes provability. Popper has shown that equipollence classes form a Boolean algebra; deductive logics also, in general, can be treated as Boolean algebras; but all that shows is the most general kind of shared structure, not at all that probable inference is a graduated generalization of deductive inference.

Cohen's argument is less direct than this. He claims that a close similarity between the types of provability regulated by differently interpreted logical systems and the various criteria of applicability of 'probable' offered by competing interpretations is demonstrated by the fact that at least five important questions which can be asked about provability—are inference rules singular or general, necessarily true or contingent, extensional or non-extensional, predicate-related or sentence-related, counterfactualizable or non-counterfactualizable?—can also be asked about probability-judgements; and 'these . . . binary dimensions of categorization provide a matrix within which all the familiarly advocated criteria of probability can be accommodated in distinct pigeon-holes' [1977: 17].

This categorization undoubtedly has an appealing explanatory neatness. But it does not commit us to pluralism. Rather, it shows that, if we accept that there is an analogy between probability and provability, then we can categorize various interpretations of probability in a manner similar to categorizing interpretations of deductive systems. However, this does not show that all the interpretations within our schema have equal—or any—worth or applicability. Some of the pigeon-holes do not have in them any traditionally accepted interpretation, and the existence of the categorization does not on its own add any inherent plausibility to pluralism. A committed monocriterial frequentist might accept that probabilities resembled proofs in the way described, recognize that a matrix of interpretations could be drawn up, yet still argue that only the pigeon-holes 'general, contingent, extensional . . .' and their contents—limiting relative frequency—are of any use as an interpretation of probability: perhaps, the frequentist would claim, these three properties are necessary for science to be able to make any use of the concept at all.

The third stage of the argument produces its most radical idea: that there is not only the internal pluralism of interpretations of the standard formal calculus but also an external pluralism—a demonstrable need for a plurality of calculi. There is a rather curious double-hit feel to this argument—as though both ends supported the middle, and each the other. The fact that generalizing provability under different inference-rules produces a lattice of interpretations of partial provability is made to suggest that most, if not all, of the holes in the lattice ought to be filled by some viable probability concept. At the same time the fact that there is a concept which can be claimed to be a probability interpretation, but does not correspond to the formal mathematical calculus, is taken as showing that probability is best seen as generalized provability. I have already argued that the forward stroke of the argument establishes nothing: to set up a categorization is not to show that each pigeon-hole must have something interesting in it. So the weight of the argument must go on to the back foot, so to speak; inductive support and judicial decisions must be shown to be probability-judgements which are, however, not conformable to the standard calculus. If this can be done, then even without the backing of the partial provability notion it will limit any account of probability which takes the calculus as given.

This latter enterprise is where the real force of pluralism, if it has any, is to be found. I shall devote part of Chapter 4 to contending that Cohen's arguments for it are inadequate. For now, let me just say that I believe that the argument about inductive support ignores three options open to the Pascalian—treating support as a matter of the *raising* of a dyadic probability, allowing non-extensionality of subjective probabilities, and interpreting weight (in the Keynesian sense) in terms of higher-order degree of belief—and that the arguments about the judicial context ignore the options of providing a consequentialist justification for distinguishing verdicts from (Pascalian) belief and, again, of developing my earlier account of evidential weight as a function of second-order probabilities.

Finally, it is worth noting that even if this third part of Cohen's grand argument were right, provided it were right independently of a partial provability thesis it would not pose any destructive threat to a subjectivist programme. Cohen's 'Baconian' probabilities resemble personalist Pascalian probabilities, he concedes [1977: 32], 'in being analogous to inference-rules that are singular, contingent and non-extensional'. Perhaps, then, a personalist might tolerate a plurality of calculi in the happy knowledge that a univocally personalist interpretation will be most appropriate for each of them.

(c) *Probability-Kinematics Pluralisms*

A very clear statement of the standard version of this view (due to Jeffrey [1965]) is given by Skyrms [1980*a*]. He argues that there are situations in which we want to distinguish 'objective' probabilities, or propensities, from our degrees of belief. One instance would be that, mentioned earlier, of a coin which may be normally two-sided or may be double-headed. As another example, suppose we know that a coin is biased two to one either in favour of heads or of tails, but we do not know which. Then, Skyrms suggests, it is natural to say that in the first case the propensity of getting heads is $\frac{2}{3}$, in the latter $\frac{1}{3}$. Our subjective degree of belief in heads would then be a weighted average of these propensities, the weights we assign depending on the information available to us about the coin. If we now come across an extra item of information we can change our degrees of belief solely by changing the weights given to the propensities—the process being called probability-kinematics. Jeffrey's contribution was to show that we can reverse the process of obtaining a degree of belief from propensities by conditionalizing out (using Bayes's theorem) to inverse probabilities, a quite legitimate move for a personalist (though not for many other schools of statisticians). He calls this reverse process 'objectification'.

Such reversals are quite sound, but that they are possible does not show the probabilities arrived at by objectification to be objective in any sense which a subjectivist would find unacceptable. Subjectivists would interpret this process as simply meaning that, from any subjective probability you have for a given event E, if you partition E into a number of constituent events then Bayes's theorem directs, for each partition, a unique value of your prior probability-assessments for the constituent events coherent with your subjective probability of E. But the fact that each partition yields unique values in no way implies that those probability evaluations must be taken to be objective. For, as Skyrms himself admits, there is no unique correct way to decompose E into constituent E_i's; so your $p(E)$ could be 'unmixed' into a large, possibly infinite, number of coherent initial distributions. If there were, in any normal sense, an objective ground for our values of $p(E_i)$ we should instead expect to find them uniquely determined. The inverse probabilities arrived by objectification are only uniquely determined within each partition of a posterior probability distribution—the choice of partition is not uniquely constrained by any objective requirements.

A more sophisticated variant on this view is due to David Lewis, who [1986] offers 'a subjectivist's guide to objective chance' which might be thought to do justice both to the realist burden and our subjectivist commitments. At least, that offering is a part of his strategically rather obscure and complex argument. As I understand it, his upbeat aim is to show that our best grip on the concept of chance is a subjectivist one; but this best analysis is developed only so that, on the downbeat, an unbridgeable gap can be pointed out between chances, thus understood, and the world. I shall argue that Lewis's account does not pose a threat to SCS because, in contrast to it, his positive account of chance is inadequate.

Lewis attempts to capture our fundamental intuitions about chance within a 'Principal Principle' which identifies chance at a world with the unique degree of belief deriving from any reasonable initial degree of belief conditionalized on the complete history and theory of chance at that world. He draws out the consequences of this principle as far as they will go towards showing chances to be supervenient on the 'Humean mosaic': 'a vast mosaic of local matters of particular fact, just one little thing and then another' [1986: ii. 111]. He offers two main modes of support for the Principal Principle: an appeal to prephilosophical intuitions and, for those untouched by the appeal, an argument from his strong commitments to viewing chance as objectified credence.

The opening sections of Lewis's paper attempt the former course, but in a manner which will sound as odd to the objectivist about probability as to the subjectivist. At least some of our beliefs are gradable—call these beliefs credences. Allow, *contra* the radical subjectivist, that in some sense of probability yet to be explained there is a kind of probability, chance, which is distinguishable from another kind, credence. Then propositions about chance will be proper subjects to have beliefs about: we will have credence about chance. Lewis now claims, *contra* the realist objectivist, that the firm opinions we hold concerning reasonable credence about chance afford the best grip we have on the concept of chance. So, it looks as if Lewis's upbeat aim will include offering an account of chance in line with these intuitive 'firm opinions' which rejects both univocal realist objectivism and radical subjectivism, while pluralistically retaining elements of both.

But what is credence *about* chance? Presumably, some credences have nothing at all to do with chance. I believe to the maximum degree of my confidence that $2 + 2 = 4$, but I take that proposition to be non-chancy. Conflicting credences in analytic propositions exist and may even be both reasonable: as when I, knowing Euclid's proof, believe to degree 100% that there is no greatest prime and you, knowing no proof, believe it to degree 50%. Aside from such cases, credences might have something to do with chance in one of two ways: either the object of the belief might be a proposition which is chancy in whatever sense '$2 + 2 = 4$' is non-chancy, or it might be a proposition which asserted something about chance, such as that the chances in the world are independent of its history. Moving to the coin-tossing example which Lewis uses to demonstrate the line our intuitions allegedly take, credence might have to do with chance either by being credence in the chancy proposition 'this coin will fall heads' or in the proposition which asserts 'the chance that this coin will fall heads is n%'. In any normal sense of 'about', we would describe the former kind as credence about the outcomes of chancy propositions, only the latter as credence about chance.

Lewis [1986: ii. 84–6] asks us to introspect as to our intuitions about the correct answers to four questions:

1. 'A certain coin is scheduled to be tossed at noon today. You are sure that this chosen coin is fair: it has a 50% chance of falling heads and a 50% chance of falling tails. You have no other relevant information. Consider the proposition that the coin tossed at noon today falls heads. To what degree would you now believe that proposition?'
   Lewis's answer: 50%.

2. 'As before, except that you have plenty of seemingly relevant evidence tending to lead you to expect that the coin will fall heads. . . . Yet you remain quite sure, despite all this evidence, that the chance of heads this time is 50%. To what degree should you believe the proposition that the coin will fall heads this time?'
   Lewis's answer: still 50%.

3. 'As before, except that now it is afternoon and you have evidence that became available after the coin was tossed at noon. Maybe you know for certain that it fell heads; . . . you remain as sure as ever that the chance of heads, just before noon, was 50%. To what degree should you believe that the coin tossed at noon fell heads?'
   Lewis's answer: not 50%, but something not far short of 100%.

4. 'Suppose you are not sure that the coin is fair. You divide your belief . . . as follows: . . . 27% that the chance of heads is 50% . . . 22% that the chance of heads is 35% . . . 51% that the chance of heads is 80%. Then to what degree should you believe that the coin falls heads?'
   Lewis's answer: (27% × 50%) + (22% × 35%) + (51% × 80%) = 62%.

Now the point of Lewis's questionnaire is to demonstrate that there is a direct link, in at least some cases, from our credences about chances to credences about the outcomes of chancy propositions. How is a realist objectivist to be persuaded of this? His response to each of questions 1–4 will differ from Lewis's:

1&4. RO: You should believe that the coin will fall heads to whatever degree there is, objectively, a chance of heads: perhaps 50% or 35%, perhaps not.

L: But there is no means of knowing what the chance is apart from your level of credence about the outcome.

RO: That's why I'm a realist!

2. RO: You should believe that the coin will fall heads to the same degree as the objective chance of heads, that is, around 90%.

L: But your credence is that the chance is 50%.

RO: So, your credence is not what it should be.

3. RO: You should believe that the coin tossed at noon fell heads. If such beliefs are gradable, you should hold this one with a large measure of confidence. But it has nothing to do with chance.

L: For very different reasons, I agree that your credence should no longer stay at 50%.

RO: But my account does not rely on a dubious notion of admissibility of evidence.

I believe the objectivist position to be untenable; but I cannot see that his responses to these questions are obviously counterintuitive. So, if Lewis's 'undefended answers' are 'evidence for what follows' [1986: 86], his questionnaire will not suffice to persuade the objectivist towards the Principal Principle.

What about the radical subjectivist? (S)he will share Lewis's intuitions about the answers to questions 1, 2, and 4 at least; but will be unable to see why these intuitions should lead to any distinction between chance and credence. Certainly, credence may be 'objectified' in the sense defined originally by Jeffrey [1965]: conditionalized upon the true member of some partitioning, some set of pairwise incompatible propositions whose disjunction is the proposition which is the object of credence. These objectified credences will be relative to a partitioning; there will be no uniquely correct choice of partitioning; 'objective probability as we have defined it here may be a highly subjective magnitude in the sense that different agents may apply the definition and emerge with different objective probabilities for a proposition A' [1965: 205]. Calling such credences 'objective' or 'chances' is in itself no threat to radical subjectivism; such a threat will emerge only if the notion of unique objectification can be made coherent without smuggling in realist assumptions. Otherwise there are just unconditional and conditional credences, with some of the latter being conditional on true propositions, others not.

The intended intuitive force of Lewis's questionnaire seems to be something like this: if, as we seem to, we sometimes hold a degree of belief m that the chance of an event is n, there must be in play two distinct senses of probability, credence or rational degree of belief as opposed to chance or propensity. But, as Skyrms [1980a: app. 2] and Jeffrey [1974] point out, it is quite compatible with our intuitions to treat m and n as just the second-order and first-order degrees of belief of an agent who does not know his or her own mind with certainty— a plausible-enough idea on a dispositional account of belief, and one which naturally leads to an account of second-order probabilities such as I developed in Chapter 2.

There is one other group of intuitions to which Lewis appeals as a counter to radical subjectivism.

The practice and the analysis of science require both concepts [sc. subjective credence and objective chance] [1986: 83]
We think that a coin about to be tossed has a certain chance of falling heads, or that a radioactive atom has a certain chance of decaying within the year, quite regardless of what anyone may think about it and quite regardless of whether there are any other such coins or atoms. [1986: 90]
counterfeit chances [sc. non-uniquely objectified chances] are therefore not the sort of thing we would want to find in our fundamental physical theories, or even in our theories of radioactive decay and the like. [1986: 121]

These assertions are not argued; rather, Lewis is saying here that we just have such pre-philosophical intuitions about chance and our interpretation of probability must respect them. Of course, though Lewis does not admit it, these intuitions rather run counter to the intuitions exhibited in the questionnaire. Suppose I know that this coin, the only one in the universe, will never be tossed. Then my degree of belief that it will fall heads is zero; so, by the Principal Principle, its chance of falling heads would be zero unless that knowledge that it will never be tossed is inadmissible, in Lewis's sense. This would be so if such knowledge bears on belief about outcomes other than via belief about chances; intuitively, that seems right. But then, does not whether the knowledge is admissible affect the chance that the coin will fall heads? How could whether knowledge is admissible be settled regardless of our beliefs about chances and of the existence of other coins—surely it is a function of our physical theories, including our theories about probability and its relation to complete and incomplete evidence?

Since intuition will not carry us as far as Lewis hopes, he must rely on the sequence of strong commitments which makes up the argument for the Principal Principle to achieve his upbeat aim of a subjectivist analysis of objective chance. Starting from the notion of chance as objectified credence, the strategy is to characterize admissibility in such a way that H_{tw} (the complete history of world w up to time t) and T_w (the complete 'theory of chance' for w) are admissible evidence upon which to conditionalize; the conjunction $H_{tw}T_w$ will be partitionable, and chances P_{tw} will be the uniquely determined objectified credences obtained by conditionalizing on the cell of the history–theory partition which is true at w and t. This analysis will carry us as far as possible towards an account of present chances as supervenient upon history up till now and the totality of history-to-chance conditionals.

This analysis is of course merely being set up to have its limitations exposed. But in any case its coherence is very doubtful. Consider first the interplay within it of the notions of resiliency and admissibility. Resiliency, as I have been developing the idea, is a measure of invariance of credence functions under further conditionalization. Such invariance is relative to some set of propositions upon which conditionalization is occurring; this must be so, because otherwise resiliency would be trivially impossible—it is always possible to form a proposition conditionalization on which will alter any credence function, no matter how resilient it has proved to be under no matter how many conditionalizations to date. Resiliency behaves as an idiolect of consensus: the invariance resembles the way in which divergent prior distributions are brought towards agreement when each is conditionalized upon agreed items of evidence. But, just as it is always possible for a new item of evidence to emerge which, conditionalized upon, will radically alter the distributions, so the invariance of highly resilient credence functions is contingent upon the set forming the basis of further conditionalization—the scope of resiliency—not containing some radically transforming item.

If objectified credences are to be uniquely determined chances, their domain of resiliency must be the maximum possible—it must be the case that no further item of evidence could emerge which would alter these credences and that no alternative partition of the domain could yield variation of credence. The former could be achieved by conditionalizing on the whole of history, or, at least, on the complete causal description of the event which is the object of the credence: the chance of the event would then be its credence 'conditional on the complete description of the whole backwards light cone of that event' (Skyrms [1980a: 45]). However, even that is not enough to satisfy the second requirement. Credences maximally conditionalized in this way could still yield different objectifications if H_{tw} remained open to different partitionings. Lewis hopes to prevent this by adding to the domain of resiliency T_w, the complete set of history-to-chance conditionals at world w [1986: 95–6]. Then, since H_{tw} and T_w together link chances uniquely to history, however the giant conjunction $H_{tw}T_w$ is partitioned, any reasonable credence conditional on the cell of it true at t and w will have the same value—the chance P_{tw}.

But the problem is that this latter manœuvre will only have this effect if it imports into the analysis a realist construal of T_w, treating it not as a theory which might be 'an incredible miscellany of unrelated propositions about chance' (Lewis [1986: 96]) but as the conjunction of all history-to-chance conditionals that *hold* at w, as the specification of the truth about chance—as Lewis immediately does [1986: 97]. This slide from all possible history-to-chance conditionals to those true at w is a slide from contingency into necessity of the relation of chance to 'ontologically complete' evidence, and thus into surreptitious realism about T_w.

That Lewis is forced into such a very obvious shuffle is an indication that his upbeat account of chance has landed him in a dilemma. This dilemma underlies the point made by Kyburg [1981] that the Principal Principle (PP) will need to be supplemented by a Principle of Integration (PI) (although Kyburg fails to see what it is that gives force to his objection—as does Lewis in his reply [1986: 117]). The criticism above can be seen as drawing attention to the difficulty Lewis's account faces as to how wide the domain of resiliency is to be for credences to become treatable as objective chances. If the domain is wide enough to include all theories T_w then credences, objectified or not, remain credences, not unique or invariant, and subject to a PP which is merely a principle of reflective self-knowledge. In this case, Lewis would be right to claim that there is no need for a supplementing PI to explain how chances are summed, because they just obey the axiom of additivity for credences. But, unfortunately, these credences are not objective chances. If, on the other hand, the domain is narrowed to those T_w conditionals which hold at w, then credences will be uniquely objectified and PP will connect credence to objective chance. But then (quite aside from the worry that this move begs the question about the reality of chance) it is no longer clear how these chances are to be combined into rational credences, since we have no principle of additivity of these chances. We will need a Kyburgian PI to connect chance back to rational judgement, and there is no route in that direction in Lewis's system.

3.3. FROM SCS TO PROJECTIVE PROBABILITY

(a) *Ontic Commitment*

Many subjectivists—chief among them de Finetti—have seen the essence of subjectivism as a triumphant wielding of Ockham's razor: thus, de Finetti opens *Theory of Probability* [1974] with the upper-case slogan 'PROBABILITY DOES NOT EXIST'. The objectivist is seen as holding that probabilities exist either as mathematical or empirical objects while the subjectivist views them as degrees of belief with the same ontological status as any opinions or beliefs. In the course of discussing laws of large numbers, de Finetti categorizes objectivist theories into those where probabilities are identified with, or derived from, mathematical properties such as symmetry and those where they stem from physical properties such as homogeneity. Both categories are, he claims, mistaken in reducing the judgements we must make about uncertain events to the circumstances we consider relevant to such evaluations. This mistaken reduction is the mistake of identifying a criterion of gradation with its means of assessment; whereas, on a subjectivist account, the probability of, say, drawing an ace, for you, is just any opinion you may legitimately hold about the outcome 'ace'. The probability of drawing an ace has the same ontological status, not as the charge of this particle or the congruence of these triangles but rather as your opinion that this particle is positively charged or these triangles are congruent.

But is there really any issue of ontology involved here? Or is it rather that there appears to be such an issue for de Finetti only because he can envisage no other form of objectivism than a naïve realism? Propensity theory may construe objective probabilities as dispositions associated with chance distributions; and it can do so without adopting an ontic commitment as to the status of dispositions. And other forms of objectivism, which treat probabilities as mathematical entities, need not be Platonist about these, or any, mathematical entities as abstract objects: a familiar point, that 'it is possible to be a realist about mathematical discourse without committing oneself to the existence of mathematical objects' (Putnam [1975: 185]). Indeed, it seems desirable in the defence of subjectivism *not* to try to sustain a rigid ontological objective–subjective distinction. For if probabilities are opinions, and opinions are subjective, and all subjective entities are ontologically radically distinct from 'real' objective entities, what other place is there for probabilities than in the mind, strongly dualistically conceived?

The picture emerging here, and from the discussion in the previous section, is that objectivist interpretations of probability need be no more committed than subjectivism to an entity-realism about probabilities. But nor is it as obvious as de Finetti takes it to be that subjectivism must oppose entity-realist commitment to the existence of probabilities. SCS as developed here has proceeded outwards from the idealized coherent individual's belief-distribution to account for the features we associate with probabilities. Its central theme is a denial that we understand statements about probability in such a way that we could be able to say that they are determinately true or not true independent of the possibility of our coming to know whether they are true. And it is at the least arguable (see Luntley [1988]) that this central tenet—whether we have a realist understanding of probability—is separable from the question whether we have an ontic commitment to probabilities, so that a negative answer to the former question is compatible with an affirmative answer to the latter.

At any rate, it is clear that the primacy standardly accorded to ontic commitment accords poorly with the methodological priorities of SCS. Coherence is, in essence, a constraint upon assertibility. So, the key to the subjectivity of subjectivism is to be found in its relation to the issues raised by semantic realism and anti-realism—issues of recognition-transcendence, of determinacy of truth-value, of which commitments possess truth-values, and so on. To oppose de Finetti's slogan let us adopt a crude capitalized slogan of our own:

THE SUBJECT IS CENTRAL TO THE SUBJECTIVITY OF SUBJECTIVISM.

(b) *Outright Anti-Realism and Epistemology*

To be a little fairer to de Finetti, I should point out that he does offer as a supplement to his ontological distinction a characterization of the subjective–objective distinction as an epistemological issue; he sees objectivism as a search for 'unknown' probabilities, whereas for the subjectivist probabilities must always be known, being the choice You make in Your state of uncertainty. But this then leads him into the worry that subjectivism in the field of probability must be defended against the charge that it naturally leads into subjectivism about knowledge in general.

A first attempt at soothing that worry would involve pointing out that anti-realism about probability-assignments is not only separable from ontic commitments (as we saw in the last section) but also separable from anti-realism about other classes of statements. Suppose we take SCS to be an 'outright', 'assertibility' anti-realism resting upon a non-objectivist semantics for probability-statements. That is, while the objectivist believes that a probability-judgement is right or wrong according as it does or does not correspond to the objective, unique probability-value, the subjectivist sees the only standards of correctness as deriving from the coherent range available for the judgement to the individual making it. Thus, the objectivist is in epistemological opposition to the subjectivist: the former is a realist about probability-ascriptions, claiming that they must be determinately true or not true even when their truth-value is, or must be, beyond our verification resources, while the subjectivist is an anti-realist in repudiating any possibility of distinguishing between a probability-ascription's being true and its meeting our best criteria of acceptability (for SCS, the single constraint of coherence). Then a fairly strong realism about our non-probabilistic statements is quite compatible with that SCS anti-realism. We can accept that there *is* a world of objects: we might turn to Hume—since de Finetti is found of claiming Hume as philosophical ancestor—for a description of such a world: it must contain beings which have a continued existence, even when not present to the senses, and which exist 'independent of the incessant revolutions which we are conscious of in ourselves' (*Treatise*, I. iv. 2). We can accept that our empirical statements must be in some way answerable to this world: again, this presents no problem for SCS. Probability-judgements are possible only because of gaps in our acquaintance with this world: when that acquaintance is enlarged, the gaps narrow—ultimately, prevision becomes impossible—and our bets are settled. We can accept a demand for epistemological objectivity requiring the existence of standards which enable people to agree on the relative worth of statements: generalized coherence supplies these standards.

Characterizing SCS as an anti-realism of this kind raises the question whether SCS is offering a reductive theory, reducing statements of the type 'It is probable to degree n that P' to 'an ideally rational individual would accept odds $n : 1 - n$ that P'. A good way to view this issue is in the light of a remark of Dummett's [1982: 85] that 'almost any anti-realist doctrine seems to lend itself to expression by means of a reductive thesis: but, in some cases, this thesis proves to be only a loose and inessential formulation of the doctrine, while in others it plays an essential rôle'. Is reduction to betting-odds for all probability attributions essential to SCS? The standard subjectivist position is that it is. But, as

I argued in §1.2, SCS is better freed from the naïve operationalism which generally motivates that reductionism; one major gain, as §2.3 showed, is that we then allow that evidence for a degree of belief can be gained other than directly by making a wager (which helps develop the theory of higher-order degrees of belief). The reduction at issue is, in any case, not an external one—SCS would not accept any reduction of probabilities to facts obtaining in the world which do not constitute part of the meaning of the probability in question. So whether an internal reductive thesis is accepted is of no great moment.

(c) *Recognition-transcendence, Bivalence, Determinacy*

The last section's presentation of SCS as anti-realism skirts a number of important issues: some of these will be tackled in the next few sections, forcing a refinement of that presentation; some I consider it possible and politic to evade or dismiss as bearing less on SCS than elsewhere.

I have until now assumed that the central question at issue in the realism versus anti-realism debate is one of recognition-transcendence; the principle of bivalence I have taken as a necessary but not sufficient condition for realism—sufficient conditions would include, at least, bivalence, recognition-transcendence, and the existence of some mechanism whereby the potentially recognition-transcendent realm determines the truth-values of statements. The debate should not be seen as focusing on bivalence alone. Recently (e.g. in Luntley [1988]), this orthodox view has been challenged: the argument is that there is an issue of objectivity-of-truth separable from that of objectivity-of-content and that the former captures the core metaphysical question of the debate. Luntley claims that anti-realism, while it must reject a realist thesis of the objectivity-of-truth, need not quarrel with realism over the objectivity-of-content thesis that 'the characterization of content requires the subject's possession of a conception of a world beyond that which is experienced'. But the objectivism–subjectivism quarrel within probability theory can and should prescind from this global argument, precisely because it is essentially a quarrel about the nature, not the truth-values, of the contents grasped in probability-judgements.

This avoidance tactic draws attention to another issue which I have so far ignored. Subjectivism is usually taken, in attacking the concept of 'unknown' probabilities, to be denying that (non-probabilistic) ground-level propositions have determinate, possibly recognition-transcendent, probability-values. Even if possession of a probability-value is best seen as a generalization of possession of a truth-value, is that position obviously the same as the one attributed to SCS—denial that probability-attribution statements have determinate, possibly recognition-transcendent, truth-values? A little argument is needed here, though not a lot. The notion of truth of probability-statements available to a subjectivist must be a very thin one. If '$p(X)\varepsilon\Gamma$' means only that 'the coherent range for $p(X)$ available to any ideally coherent agent is a subset of Γ', prefacing the former statement with 'it is true that . . .' is thin enough to be transparent as to meaning. So a semantic anti-realist expression of SCS is bound to cover any metaphysical anti-realism expressed by it about statements of '$p(X)\varepsilon\Gamma$' form.

(d) *Truth-conditions, Commitment, Projection*

The line of argument just pursued suggests an instrumentalist or 'non-cognitivist' shape to an anti-realist SCS. Perhaps probability-judgements do not have truth-values at all. Consider the archetypal subjectivist betting scenario. Suppose two agents A and B are considering an event E, uncertain to both their judgements. Then, for A to assert that $p(E) = a$ is merely to say that (s)he will accept odds of $a : 1 - a$; similarly for B to assert that $p(E) = b$ is simply commitment to the odds $b : 1 - b$. There can be no contradiction between the two judgements, even if a differs from b, because the statements are not truth-bearers; the conflict between them is that between two competing evaluations or non-assertoric speech acts. Once the outcome of E is known to both A and B it ceases to become an object of previsions; statements about it cannot ascribe probability to it, and can bear truth-values—so a subjectivist may well be realist about these statements.

Certainly, SCS could adopt such an instrumentalist form of anti-realism. But it now seems to me preferable to be wary of presuming a clear-cut distinction between beliefs, which bear truth-values, and other forms of commitment—call them stances—which do not. To do so is to remain in the grip of too realist a conception of truth, or at least of a conception of truth which SCS does not need. It is legitimate, and desirable, to start from expecting that some commitments fall naturally into a truth-conditional form of expression, others not; there are commitments which expose one to, for example, loss—or perhaps sin— instead of defeasibility. What is dubious is the assumption that we must end up with an absolute distinction.

What is emerging here is an interpretation of SCS as a theory of projective probabilities, of subjective probabilities as projected personal evaluations. But it is not a relativism about probability—not open to the old gibe at subjectivism that it is just the claim that 'sincerity is enough'. For, as Chapter 2 has shown, the coherence constraint, understood dynamically, is sufficient to provide an assertoric standard which makes sense of the idea of improvements on evaluation-distributions and which compels their convergence. SCS could choose an instrumentalist line, dispensing with truth in favour of warrant. But it does better to undertake a quasi-realist enterprise of showing how, starting from a view of probabilities as stances, we distil from minimal rationality standards a notion of truth indistinguishable from that emerging from a more florid realist metaphysics. That is the positive project which SCS should best be seen as undertaking.

(e) *Surface Realism*

Alongside that positive project, SCS must take up the burden of surface realism—must account, via projectivism, for realist-sounding probability practice. This task will intertwine with the positive project of the last section; for now, let me indicate briefly two elements of that practice (others have of course been implicit in earlier discussion).

1. A standard argument for a realist construal of truth is that, while we are prepared to say that some now-falsified hypothesis was accepted by, or was rationally acceptable to, or met all the best assertibility criteria of, some group or individual at a time before it was falsified, we do not say that it was true—rather that it *seemed* true. Similarly, we say that a hypothesis seemed rather than was probable. Now, as I pointed out in §3.2, it is implausible to use this point in order to assert a radical meaning-variance between 'probable' applied to hypotheses and 'probable' applied to events. But the point could support a difference in supervenience conditions. On various different ways of viewing probability, it is natural to regard the probability of an event as supervenient upon events in the world—which do not include possession of, or lack of, items of evidence. That is, the probability of an event cannot alter unless some event-like particular alters; but the probability of a proposition would seem to be affected just by the acquisition of evidence relevant to the proposition. This point seems to impel us towards some sort of distinction between objective chance and subjective credence. But, as the discussion of Lewis's system has shown, it can be accounted

for by a quasi-realist SCS rather than driving us into realist objectivism or pluralism.

2. L. J. Cohen [1986*a*] makes out a powerful case to the effect that people are naturally inclined to think in terms of counterfactualizable rather than non-counterfactualizable probabilities wherever possible, reflecting a general dominance of causal—or as near as possible to causal—reasoning over reliance on bare statistics. This well-supported psychological hypothesis is difficult to explain within any univocal interpretation of probability: *a fortiori*, within SCS. I shall not attempt a full answer to this argument, but §4.1 will indicate some possible responses, which derive from the quasi-realist freedom to offer differing analyses of our projectivist tendencies in assigning probabilities and our projectivist tendencies in relying upon induction.

(f) *Projective Probability*

§1.4 and §2.3 discussed the challenge posed to SCS by calibration paradoxes and suggested that that challenge could be met by distinguishing the rationality of probability-judgements from their vindication. If this notion is now developed further by connecting it with the notion of resiliency introduced in §2.4 and with an analysis of higher-order probabilities, it will be possible to see how the anti-realist starting-point of SCS gives rise to a quasi-realist achievement.

Let us recall that the calibration theorem appears to demand a totally implausible level of confidence that one is perfectly well calibrated, given only that one is coherent. Coherence, however, provides no guarantee that your judgements will be vindicated, only a guarantee that they are not guaranteed not to be vindicated (by your being open to a Dutch book). It is rather a constraint upon unbridled irrationality than a limit of rationality.

The link between rationality and vindication lies in this: the paradigm of irrationality is to form, organize, or change your partial beliefs in such a way that you needlessly hinder your chance of vindication—in the worst case, guarantee that you cannot be vindicated. Coherence of your probability distribution renders every possible combination of your beliefs capable of vindication. You will be well calibrated if every possible combination of your beliefs actually is vindicated (in the long run). The calibration theorem shows that coherence entails that within your probability distribution should be a belief, to which you have the strongest possible commitment (probability 1), that you are well calibrated—a belief that every possible combination of your beliefs will be vindicated. From any external viewpoint, this belief will be judged to be almost certain not to be vindicated—but from no such viewpoint, nor within your distribution, is it irrational: it is not guaranteed not to be vindicated. Any other belief you held about your own calibration would be irrational, since it would guarantee that some combination of your bets could not be vindicated. So total self-confidence is entirely rational and indeed rationally compelled by coherence.

But this rational compulsion does not unreasonably rule out non-sceptical doubts. A third party can rationally, of course, doubt that any or all of your first- or higher-order beliefs will be vindicated or are rational. Given that you are coherent, you are by definition not irrational—no combination of your beliefs is guaranteed not to be vindicated. Further, because of the calibration theorem, you cannot doubt that every possible combination of your first-order beliefs will be vindicated. But it is only those first-order beliefs which are calibrated, that is, compared with relative frequencies. Coherence does not compel you to believe (at a third level) that your second-order assignment of probability 1 that you are well calibrated will be vindicated. You must, to be coherent, be entirely self-confident, but you need have no particular level of confidence that that self-confidence is justified.

That, I think, shows how one might dispose of the calibration paradox. One cost—or, from another angle, one benefit—of this analysis has been to uncover a deeper problem about coherence. Any Dutch-book-avoiding bettor is rational in the limited sense which, we have seen, is all coherence asks; but (s)he need not be rational in any everyday use of the term. Strong subjectivism demonstrates that coherence implies conditionalization which in turn implies convergence of opinion, but it does not seem to contain within itself any reason to prefer opinions conditionalized upon a great deal of evidence to those, no matter how extreme, which have not yet been much conditionalized. It offers no reason to prefer to judgements which are, we might say, on a rational track those which have gone further along that track. This surely runs counter to our normal understanding of rationality: we will not regard someone who proposes to travel from Oxford to London on a route via Saturn and Pluto as just as reasonable as someone who takes the direct road, even if we can see that London must lie on the proposed interplanetary route. Then how is this preference for evidence-increase to be explained without lapsing from subjectivism into some form of realism? Is this in fact an obstacle in the way of a successful quasi-realist subjectivism?

Consider how far that programme—operating with the single constraint of coherence—has reached in founding apparently realist practice on a much more austere metaphysics. It begins by trusting our anti-realist instincts in that it accords primacy to constraints upon warranted confidence internal to some agent. What is fundamental to defining probability is coherence, a rational recipe for an agent to avoid guaranteed non-vindication of beliefs. The representation theorems of de Finetti and Savage establish that the rational use of this recipe leads to stable and shared sets of judgements in many cases, without making such sharedness or stability an 'objective' limitation on rationality in other cases. Features of the world—frequency, symmetry, etc.—which on a realist account *determine* probability precipitate out of this quasi-realist account: the subjectivist can do with them all an objectivist can, but is not restricted by realist methodological priorities. Calibration challenged this programme by threatening to impose immediate realist consequences onto the primary notion of coherence. This threat could be disposed of (though only so) by taking the quasi-realist programme further back into the concept of coherence, identifying the rationality of partial beliefs as independent of their vindication. The alleged irrational self-confidence generated by the calibration result can then be seen to be perfectly rational—as are our doubts about it.

This response to the problem can be made more precise by utilizing the theory of the link between *resiliency* of first-order probabilities and the *values* of second-order probabilities which I developed earlier and in my [1991]. I argued that one plausible approach is to take $p_2\{p_1 (r) = \alpha\} = k \times \beta \times \text{res } p_1 (r)$'s being α, over the fixed domain, for all α. (The constant of proportionality k here is determined by the coherence constraint upon p_2 which requires that the sum of all the permissible p_2-values be 1.)

In most decision situations this suggestion will not produce precise p_2-values. The cases where greater precision *is* possible are those where the domain of resiliency comprises linear combinations of exchangeable first-order propositions. This includes the contexts to which calibration applies, where we are offering previsions about classes of exchangeable events and conditionalizing over a domain which includes feedback information about the fit between the previsions and the outcomes. In such cases, the convergence results described earlier amount to the claim that, where a law of large numbers applies, as conditionalization is based on more and more evidence, for exchangeable events, the probability that the probability of the hypothesis's truth will exceed

α approaches 1, for all $\alpha < 1$. Generalized, as indicated earlier, this becomes: increasing resiliency, given first-order exchangeability, tends to concentrate p_2 onto fewer p_1-values and the p_2 of these values approaches as close as coherence permits to 1. (Normally p_2 will concentrate onto a value near 1 attaching to one p_1-value.) It follows that, in the limit, the perfect resiliency of our first-order judgements will concentrate our second-order confidence value onto 1; but our third-order distribution (if we can have such) will not be affected.

But when we then go on to ask whether and why such p_2 concentration (equivalent to high p_1 resiliency) is rationally to be preferred, does this not take us into an area where we are under some rationality constraint other than coherence? Blackburn [1980: 195], for instance, has argued that 'standards for proper projection . . . go beyond coherence and . . . dynamic coherence . . . they involve a proper respect for frequencies, involve a proper respect for induction'. The question whether this one remaining gap in the SCS quasi-realist enterprise represents a rebuttal of SCS's ambitions or a limitation on them or neither is, in effect, the question whether the projectivism about probability which I have been advocating is committed to a claim that it obviates the need for independent standards of inductive projection or confirmational commitment. I turn to this issue in the next chapter.

4

Some Applications of the Projectivist Theory

THIS chapter considers some of the implications, for some wider issues, of a projectivist theory of probability.

The first two sections advocate caution as to the scope of the theory's application.

§ 4.1 looks at, and rejects, the claim that Bayesian convergence mechanisms alone dispose of the problems of induction; to focus the discussion, I consider those claims principally as exemplified by Dorling in a [1981] symposium with Miller.

§4.2 looks at, and rejects, the claims of zealously Bayesian philosophers of science that strong subjectivism is adequate of itself to account for theory choice or, indeed, scientific method generally: a position most recently exemplified in Howson and Urbach [1989].

The last two sections deal with areas where the theory can, I believe, be applied to resolve outstanding problems.

§4.3 explores the links between probability, explanation, and conditionals; I advocate a refined version of a statistical-relevance contrast-class account of explanation, argue that many conditionals are best seen as condensed explanations, show how these claims permit the resolution of various parallel difficulties for adequate accounts of explanations and of conditionals, and point up the role of probability within these accounts.

§4.4 turns to questions of juridical probability and argues that PP enables us to meet five apparent problems about standards of proof without being forced into the Baconianism advocated by L. J. Cohen [1977].

4.1. PROJECTIVE PROBABILITY AND INDUCTION

Enthusiastic proponents of a subjectivist interpretation of probability have often claimed that its account of consensus arising out of Bayesian conditionalization will dissolve the practical and sceptical problems of induction. Thus, de Finetti writes [1974: ii. 232]: 'In our formulation, the problem of induction is, in fact, no longer a problem . . . Everything reduces to the notion of conditional probability.' And Dorling has argued [Dorling and Miller 1981] that the theory provides us with a rational recipe for modifying our pre-existing expectations about the future in the light of incoming evidence—in other words, a direct answer, granted only that we do not expect inductive inference to yield certain knowledge, to the practical problem of induction. With the Bayesian method in our armoury, he claims (less confidently), the traditional sceptical arguments are bound also to strike us as misguided.

It is worth pointing out, to begin with, that the distinction being presumed here between 'practical' and 'sceptical' problems of induction is, despite its widespread currency, by no means obviously easy to sustain. A clear and relatively early formulation of this distinction appears in Levi [1967], where the global issue of induction is understood as the (hopeless) effort to solve Hume's problem, in contrast to the local issue of making sense of the inductive arguments people actually employ within some context of enquiry. Questions of inductive justification, Levi argues, arise only as questions about the propriety of a particular revision of a shared rational corpus of beliefs—and some conditionalization principle can be expected to settle such questions. But the problem with such a view is that its assumption that everyone facing a real problem shares a large body of knowledge can have no more than a pragmatic grounding in an arbitrary setting of limits to human disagreement. It is just such an assumption which the sceptic challenges in demanding standards of justification which will apply given an arbitrarily small shared corpus: just that is the point of rule-following scepticism or of Goodman's paradox. There is then no non-question-begging way of holding the local line, of preventing local issues from being pressed, via insistence that we have here no a priori canons of rationality, until they become global.

Be that as it may, one may concede that Dorling has assembled a strong case for supposing that several of the main standard sceptical attacks—counterinductivism, Goodman's arguments, and millenarian scepticism—threaten a Bayesian conditionalization account less than they do any justificationist analysis of induction.

Counterinductive scepticism fails because it relies on a symmetry which Dutch book considerations free the Bayesian from having to adopt. Goodman's paradox loses its bite because either it must rely upon an equality of prior probabilities as to, say, observed emeralds at some future date becoming identified as grue or as green—which equality necessitates a symmetry to which the Bayesian is, again, not committed; or it will reduce to millenarianism. Finally, millenarianism fails because the Bayesian is perfectly able to tolerate the existence of the millenarian (if, as is most unlikely, the latter can remain coherent), since his or her position will not threaten other coherent sets of previsions. However, his greater confidence that there is a quick and easy Bayesian answer to the 'local' problem is obviously misplaced. In his reply to Dorling [Dorling and Miller 1981], Miller argues that conditionalization lacks the quality of unique compulsion on our judgements necessary if it is to guide statistical or scientific inference. This is a point made forcibly also by Kyburg [1978a: 176]:

> there is nothing in the theory which says that a person should change his beliefs in response to evidence in accordance with Bayes' theorem . . . Just as he may have got to his original coherent state by intuition, whimsey, imagination, evidence processed through Bayes' theorem, or any combination thereof, so he may with perfect rationality pass from his present coherent state to a future coherent state through any of these mechanisms. If he depends on Bayes' theorem, it is a matter of predilection, not of rationality. For all the subjectivistic theory has to say, he may with equal justification pass from one coherent state to another by free association, reading tea-leaves or consulting his parrot.

As subjectivism is presented by Dorling, there is no answer to this charge; indeed, as Miller is quick to point out, Dorling effectively [Dorling and Miller 1981: 122] concedes it.

The theory developed in Chapter 2 suggests an answer which might seem to restore the Bayesian claim to provide a rational recipe for inductive inference: contrary to Kyburg's assertion, there is something in subjectivism which says that we must conform to Bayes's theorem, and that something is exchangeability. As we have seen, if you have a coherent probability distribution over non-exchangeable events, then, while applying Bayes's theorem to your conditional probabilities as new evidence is acquired will ensure that you remain coherent, there could well be other coherent distributions which you could arrive at by some other means—so the criticism will hold here (at any rate in theory: in practice, coherence is so tight a constraint that this is extremely implausible). But if a class of events is exchangeable as far as your beliefs are concerned, then it follows from the representation theorem that the coherence requirement alone generates a unique probability density function over those events; when we are dealing with conditional probabilities this unique function is precisely that generated by conditionalizing in accordance with Bayes's theorem. And, in cases where frequency information is available, this posterior distribution must correspond to the observed frequencies. So Bayesian conditionalization is 'compulsory', though only for classes of exchangeable events; in effect, exchangeability in time-heterogeneous classes—the most usual case of interest to science or to statistical inference—serves as a refined version of a guarantee that the future will resemble the past.

This is undoubtedly an improvement on the account which Dorling offers; but is it adequate to sustain a claim that the problem of induction reduces to that of the revision of our partial beliefs? There are strong reasons to suppose that it cannot be adequate. As I pointed out earlier, Jonathan Cohen (beginning with his [1970] and in many places since) has drawn attention to a crucial problem for any enterprise setting out to show that inductive support-functions can be identified with probabilities or mapped onto them. Beginning from the weak assumptions that the value of an inductive support-function ought to remain invariant if the argument is replaced by a logically equivalent proposition and that any inequality of inductive support for two generalizations is possible only if there is an inequality of inductive support for their substitution-instances, he formulates what he calls the 'instantial conjunction principle': that the support offered by any body of evidence to any conjunction of instances of a generalization must be the same as the support offered to any single instance. Consequently, it proves impossible to identify inductive support-functions with probabilities, since there is a direct conflict between this principle and the multiplication law of probabilities. Moreover, such Bayesian enthusiasm overlooks much which an adequate account of inductive practice ought to include and which the conditionalization model certainly does not: the value of variety of evidence, questions of relevance, the relation of theory to observable evidence, and so on. More importantly, it mistakes the fact that de Finetti's representation theorems create a central place for frequencies within a subjectivist model for a justification of our inductive habits as arising out of no other constraint than coherence.

The analysis which I have developed here should, I hope, demonstrate that such enthusiasm is misplaced. Coherence requires conditionalization; conditionalization yields highly resilient distributions in exchangeable contexts and there concentrates second-order probability onto observed frequency. But, as we have seen, even to get as far as accounting for our respect for weight of evidence without threatening to bring a wholly independent parameter to bear on decision has required a number of assumptions which are plausible in a description of our inductive habits but in no way displace induction from its central place in our inferences. The generalization of coherence I offered allowed me to advocate as plausible taking p_2-values as functions of resiliencies of p_1's—but not to establish it as compelling; and, crucially, we cannot avoid in characterizing our preferences for better-evidenced judgements introducing a w-factor representing our judgement of how little resiliencies might alter if the untested scope were to be tested. The inductive habit is just the preference for as high a value of w as possible, achieved by increasing the domain size and/or, as far as we can pragmatically determine, ensuring the absence from the scope of factors causally relevant to the conditionalizations. My account here offers standards of proper projection, emerging from coherence alone, which account for rational consensus in probability assessment. But it does not, and does not attempt to, *justify* the inductive habit, the preference for evidence increase. What it does do is to show that weight of evidence can be accounted for as a function of Pascalian probabilities and of w, so that understanding and justifying our respect for evidence does not require the introduction of non-Pascalian measures of belief but does demand a justification of our w-preferences: which is precisely the problem of induction.

Nothing I have said here sets limits to the form such a justification might take, whether anti-sceptical, naturalist, or whatever (and there are, as I have mentioned, alternatives to this weight-based treatment which a subjectivist might explore). Nor does it constrain the analysis of our w-preference criteria: it may be perfectly possible to transpose into it, more or less wholesale, Cohen's Baconianism, method of relevant variables, gradation of legisimilitude, and all. None of that need threaten the claim that SCS provides a complete univocal interpretation of probability and that all probabilities are Pascalian. Cohen's arguments demonstrate that the whole-hog Bayesian approach to induction is mistaken; a mistake, I believe, arising out of an over-zealous effort at protecting the univocal character of subjectivism. But it would be equally over-zealous in the opposite direction to extend a non-probabilistic account of induction into an attack on the claim that Pascalian probabilities fill the whole space of judgements which can properly be termed 'probable'.

An adequate subjectivist theory of probability must account for how internal rationality constraints on projected partial beliefs issue forth in apparently external constraints on probabilities: that is the quasi-realist task which I have been attempting. But it would be altogether too optimistic, as well as too demanding, to suppose that this is the same task as that of offering a projectivist account of our inductive expectations. This latter task, I would claim (following Ellis [1979]), has at its centre the need to develop rules of inductive projection which emerge from constraints upon our theoretical involvement with the world, not just from constraints upon our partial beliefs about particulars.

4.2. PROJECTIVE PROBABILITY AND SCIENTIFIC BAYESIANISM

If, contrary to the arguments of the preceding section, a complete subjectivist account of probability were of itself a resolution of the local problem of induction, it might be tempting to look to subjectivism as the basis for a refined positivist account of scientific method and reasoning. Thus, Dorling [Dorling and Miller 1981: 113]:

Bayesian conditionalisation is the single unifying fundamental principle underlying all legitimate forms of statistical inference. (It is still daring to maintain this also for all legitimate forms of *scientific* inference, but some of us are prepared to assert this, and to claim that detailed investigation of case-studies from the history of science lends cogent support to such a position.)

Despite their considerable differences, Salmon [1967], Rosenkrantz [1977], and Horwich [1982], among others, share this extended Bayesian enthusiasm. Most recently, this view has been very forcefully advocated by Howson and Urbach [1989], who make the strong claim that all theory choice is essentially a matter of comparison of probability in the light of evidence.

There are, of course, weaker claims which might be advanced in this context which are more or less uncontentious, if not very interesting. Probability theory surely does bear on some choices between statistical hypotheses within larger scientific theories; Bayes's theorem surely has applications to such cases; and, no doubt, probabilities and utilities are involved in most or all decision-theoretic comparisons of the costs and benefits of scientific research programmes. But altogether different and, as Dorling says, 'daring', is the adoption of the stronger position, the core of which has come to be called 'scientific Bayesianism'—a yielding to the temptation to rely on the successes of probabilistic subjectivism for solutions to long-standing problems of scientific method.

The strength of this temptation should not be underestimated. Somewhere near the centre of any philosophy of science must come the questions of how scientific disagreements are, or should be, resolved, how such a resolution relates to truth, and, consequently, to what extent the prime aim of science should be seen as achieving or approaching truth or increasing verisimilitude. One very attractive response would be to adopt a pragmatic view of the nature of scientific truth: assuming or establishing that scientific investigation, if carried out by a community of investigators relying upon methods justified independently of claims to truth, will result in the end in agreement across that community, we might then define a scientifically true proposition as one on which all such reasonable investigators will eventually agree. It then becomes very appealing to look to the convergence mechanisms of Chapter 2—particularly when applied in exchangeable contexts—to provide that independent justification of scientific method. On such an account, any hypothesis not yet contradicted by the evidence must be allowed (though it cannot be required) to have a positive probability subject only to the constraints of coherence. As evidence accrues, initially different probability distributions converge, with the consequent elimination of some of the competing hypotheses from the coherent range. If at a particular stage in this process several conflicting hypotheses remain coherently tenable, then the one with the highest monadic probability at that time is to be preferred. Exchangeability will account for the role of experiment, and the importance of frequency observations, in the process; the 'subjective', agent-relative element will be minimized by the tightness of the coherence constraint; conditionalization will account for the fact, yet the fluidity, of interim consensus.

Such Bayesian accounts of scientific method are, however, open to at least five quite different kinds of powerful objection. It may be possible to find answers to the first two of these, but the last three seem to me quite destructive of the scientific Bayesian programme.

First, there is the difficulty that subjectivism must leave enquirers' possible probability distributions prior to any empirical investigations entirely open (subject only to coherence). Some Bayesians—and even some of their critics—have represented this as simply a form of open-mindedness and tolerance entirely appropriate to science. But that response is much too sanguine. Suppose I am a fanatic about some area which we are about to investigate together scientifically. As a fanatic,

I hold, by the standards of everyone else, an extreme prior probability distribution. As a canny fanatic, I can see that conditionalization on the evidence I foresee coming in would being me into, or near to, consensus with everyone else. So, I simply shift my initial distribution to an even more extreme one—I become a wild-eyed fanatic—in order to effect unlimited delay in such consensus. Is there a Bayesian riposte to this? Perhaps: either through ruling against the insincerity of such manœuvres; or through taking Bayesianism to include an account of actual widespread consensus, not some absolute prescription against wilful deviance. But neither move is straightforwardly a merit of the Bayesian programme.

Secondly, there is the problem of how universal generalizations may, within subjectivism, have probabilities assigned to them: if the central measure of partial belief is betting-behaviour, how can open-ended hypotheses be the subject of bets—how could such bets be settled? Earlier, I listed a number of possible answers to this question, none of which—with one possible exception—could however be adequate to sustain the account just given of hypothesis choice. The only possible line, I suggested, would be to attempt to define the probability of a generalization by recourse to the notion of exchangeability. So, an as yet unfalsified generalization such as 'all swans are white' reduces to an infinite conjunction of propositions, some known for certain, some about which we are uncertain: we are entitled to exercise our prevision about each of these uncertain events provided those previsions can be made coherent. Since it is characteristic of scientific generalizations that they are indifferent to the order of our observations we may regard these constituent events as exchangeable, and so it would follow that our subjective probabilities would concur with observations of statistical regularities among these events. So it would seem that we could form a probability evaluation of the truth of the hypothesis as a combination of the certain knowledge of those elementary events known to have occurred and the probabilities of the other elementary events which we preview in line with our knowledge of frequencies. But this approach only seems to work because, by stipulating that the generalization be unfalsified, we avoid—in a way which we cannot in general achieve in scientific investigation—the consequences of the asymmetry which normally exists between verifiability and falsifiability. If, for example, you bet someone that there are no days on which men turn into rhinoceroses, you could lose this bet but could never win; and that appears to be an irrational way to bet, at any odds. So, we need some supplementary rules which Bayesianism lacks.

An alternative approach has been suggested by Howson and Urbach: to treat all partial beliefs as measurable by the betting-quotients an ideally rational agent would choose on the *possibly counterfactual* assumption that the bets can be settled—bets on generalizations being one case where the assumption *is* counterfactual. However, the problem with this suggestion is that there seems to be no guarantee that it will generate a unique betting-quotient. I would like to believe that it is likely my colleagues will all be sorry when I am dead and gone. How likely? Betting here seems pointless, if not meaningless: I shall not be around to collect. I *could* make the counterfactual assumption that I will be able to collect—but I could envisage such collection very differently in differing circumstances, and that will affect the odds I am prepared to accept. For instance, suppose the currency of our betting is to be cool beers. One cannot use the ratio of the number of beers I am prepared to stand my colleagues now to the number they wil provide me with, on the assumption they could do so after my death, if they felt sorry, as a measure of the strength of my belief that they will be sorry: for cool beer wil be immensely more valuable to me if I find myself at that time in one of hell's more torrid rather than more frigid zones. Depending on which circle of hell I expect to occupy, it would be rational for me to accept very different betting-odds.

Thirdly, even if these problems can be overcome, so that open-ended hypotheses can have subjective probabilities, the discussion of Chapter 2 and §4.1 has indicated that there is a fatal ambiguity in the Bayesian prescription for choosing between rival hypotheses. For there are two possible bases for comparison of hypotheses: credence (current subjective probability) and evidential weight. Now suppose that at some time we are considering two contrary hypotheses, A, which has high probability but has undergone relatively little conditionalization, and B, which has a lower probability but greater evidential weight. Peircean optimism as to convergence convinces us that in the end, when all the evidence is in—or, even more optimistically, probably when just a reasonable amount of the evidence is in—one of these hypotheses will have dropped out of the coherent range of possibilities. But how does that help us now? How can we anticipate what will happen to the conditionalized probabilities of A and B as that distant goal is approached? Can we be sure that B will remain less probable than A throughout? Indeed, one might plausibly claim that in the early days after a radical paradigm shift in a special science it is not uncommon for an hypothesis to be preferred to both A and B despite at the time having a lower probability and a lower evidential weight than either *A* or *B*.

Finally, just as with attempts to reduce induction to conditionalization, a Bayesianism which ignores the elements in theory formation which go beyond predictive success cannot be adequate to represent—much less explain—the full range of scientific activity. It must miss out at least two central aspects of the formation of scientific theories.

(*a*) There is the problem of how to restrict, without eliminating entirely, *ad hoc* revision of theories. Consider, for example, the impact upon conflict between creationist and natural selection theories of the origin of species of evidence about the geographical distribution of similar species. Darwin [1859] frequently draws attention to what he calls 'the most striking and important fact for us in regard to the inhabitants of islands . . . their affinity to those of the mainland, without actually being the same species' [1859: 386], arguing that 'this grand fact can receive no sort of explanation on the ordinary view of independent creation'. Following Richard Miller [1987] in reconstructing this argument in a Bayesian style, we may suppose that our prior probability for individual creation is at least as high as for natural selection; our prior dyadic probability that island species will more resemble species on the nearest mainland than species in similar but geographically remoter environments, given the truth of individual creation, is very low, but given the truth of natural selection, is very high. We then observe that geographical proximity dominates environmental similarity in the matter of species resemblance. Applying Bayes's theorem will now yield a higher posterior probability for natural selection than for individual creation. But the problem is that it need do so only if the likelihood ratios for each hypothesis (that is, the ratios of the dyadic probability of each hypothesis given the evidence to the monadic probability of the evidence) are either fixed or in some other way guaranteed to be themselves in a fixed ratio. Otherwise, if *ad hoc* revisions are permitted to one's assessment of the evidence—say, concluding in the example case that the environments on island and mainland must after all be similar in the respects needed to make species resemblance the most appropriate choice for a Creator—one can always preserve the initial probability distribution by revising the likelihood ratios. Bayesianism hence provides absolutely no rationale for resolving such disputes unless it is supplemented by a general rule prohibiting such manœuvres. A tolerant rule would make it much too easy for pseudo-science to save itself in the face of falsifying evidence. But any general prohibition would be much too restrictive, since it would mean that any theory tolerating anomalies would fall foul of it—including Darwin's own and many generally accepted scientific theories in other fields.

(*b*) Bayesianism also has difficulty in accounting for the very existence of scientific theories which are more than mere statements of observed regularities. We can always construct a phenomenological rival to any theory, a ghostly *doppelgänger* which states all the well-established observations contained in the original theory while expressing total agnosticism about the causal factors, if any, responsible for the phenomena. Bayesian hypothesis choice can never favour the causal theory over its phenomenological rival, because there can be no bets with determinable outcomes which could be won by an adherent to the former which would not also be won by an adherent to the latter. Consequently, there can be no situations in which the causal theory is vindicated at the expense of the phenomenological theory. For a thoroughgoing operationalist and instrumentalist like de Finetti, of course, this is no problem: but it is for those more recent and more reticent Bayesians who advocate their approach as possessing the advantages of those earlier positions while avoiding implausibly strong anti-explanatory commitments.

In the end, what is important is to see that the success of a projective theory of probability does not stand or fall with the plausibility of the claim that it provides any sort of adequate basis for a full-blown reconstruction of scientific methodology. Here, as in inductive logic and decision theory, the unfortunate appropriation of the term 'Bayesianism' to wave as a party banner does much to obscure these crucial distinctions.

4.3. PROJECTIVE PROBABILITY, EXPLANATION, AND CONDITIONALS

Part of the argument of the previous section has been that attempts at a Bayesian theory of scientific method generally fail because there is nothing, within subjectivism itself, to account for the role of explanation in scientific theorizing. And part of the argument of Chapter 2 has been that standard subjectivist accounts of conditional probability which identify it with the probability of conditionals fail when they encounter Lewis's triviality result or L. J. Cohen's contraposibility paradoxes. So, a projectivist theory cannot replace explanation by conditionalization or reduce the probability of a conditional to a dyadic probability. But it should, and can, have something to say about the nature of explanation and (at least some species of) conditionals, the relationship between them, and between each of them and probability. It should do so, in the case of explanation, because only a theory based on probabilistic relevance has any hope of explaining explanation. It should do so, in the case of conditionals, because even if one can reconstruct convergence mechanisms without basing them on problematic notions of the probability of conditionals, it remains true that we do have such notions and they are problematic. To quote Price's [1986] example, one often seems to have fairly clear-cut degrees of belief in conditionals such as 'if it is raining in Moscow, the Kremlin roof is wet' while having no commitment to any particular level of absolute credence in either antecedent or consequent.

That the theory of probability which I have been advocating can play an important role in our understanding of these problem areas is what this section will set out to show. The account I offer makes use of recent attempts (e.g. van Fraassen [1980a], Gärdenfors [1988]) to treat explanations as context-dependent answers to why-questions, and takes over from Gärdenfors the ideas that there can be degrees of explanation and that explanations resemble conditionals in that both involve contractions of an agent's epistemic state. However, I go beyond the former tradition in, rather than taking contextual factors as pragmatic aspects of the explanatory process, arguing that a proper understanding of the contextuality of explanation requires paying attention to the structure of explanada in a manner which also illuminates the construction of an adequate semantics for counterfactual conditionals. And, unlike Gärdenfors, I restrict myself to claiming that my analysis casts some light on the role probability plays in some aspects of explanation and conditionals, rather than embedding this analysis in an all-embracing theory of the revision of non-probabilistic as well as probabilistic belief-systems. I begin, in (a), by stating and arguing—in rather a condensed form—for the view that explanation typically consists in offering context-dependent and probability-based answers to questions of the form 'Why P rather than its contrary Q?' or 'Why P rather than a class of contraries $Q_1 \ldots Q_n$?'. This account of explanation will be embedded in a statistical-relevance rather than a deductive-nomological approach, though I shall not attempt here to justify that preference. I explain how a contrast-class theory of explanation enables us to deal with problems about reference-class choice, about whether statistical relevance is sufficient for explanation, and about asymmetry in explanations. I conclude that the account suggests and permits the view that there can be degrees of explanation.

In (*b*), I reject one popular account of the connections between explanations and conditionals in favour of treating conditionals as condensed explanations with structure 'If T_p, then P rather than Q'; I deal with a number of obvious objections to this radical suggestion.

In (*c*), I examine the way in which such a view of conditionals enables us to tackle three well-known groups of problems, which can be seen as parallel to the difficulties for accounts of explanation dealt with in (*a*). I conclude that, in general, the probability-value $p(A \rightarrow B)$ measures the explanatory power of A in a belief set containing both A and B over which that conditional connective $\rightarrow$ has been defined.

(a) *Explanations, Contrast-Classes, and Probabilities*

Theories of explanation which treat explanations as arguments in which a set of covering laws and initial conditions entails, or bestows a high probability upon, the explanandum run into great difficulties (see Salmon *et al.* [1971]). Many of these can be overcome by adopting instead the view introduced by Salmon that an explanation is an assembly of facts A_i statistically (i.e. probabilistically) relevant to a phenomenon E—that is, for each i, $p(E|A_i) \neq p(E)$. But at least three problems remain. First, it seems that defining statistical relevance for any E requires us to consider E as a member of some reference-class: how is that class to be chosen? Secondly, several examples (due to Cartwright [1979]) seem to indicate that statistical relevance is not sufficient for explanation. Thirdly, the theory has difficulty over cases of asymmetry of explanation, where two propositions are statistically relevant to one another, but one can be used to explain the other, not vice versa. Salmon has more recently [1978] tried to meet these difficulties by importing causal processes and nets of interactions into the notion of relevance: explanations are exhibitions of the relevant part of the causal net leading up to the explanandum. However, there are other problems introduced by making causal processes so central: what of laws of coexistence?; or quantum-mechanical events where there is no continuous spatio-temporal 'net'? An alternative and a better way of handling the three difficulties is to retain the non-causal statistical-relevance account, but add to it an analysis of the explanandum as well as the explanans. Let me first state this position, then apply it to these problems.

The germ of this idea—attributed by van Fraassen to Hansson [1975]—is that why-questions are systematically ambiguous. The question:

Why did Adam eat the apple?

can be construed in various ways, such as:

Why was it *Adam* who ate the apple?
Why was it *the apple* Adam ate?
Why did Adam *eat* the apple?

The underlying structure of a why-question is, then,

Why *P* in contrast to (other members of) *X*?

Generally, the contrast-class *X* is not explicitly specified because it is clear to both speaker and auditor what class is intended. One immediate point in support of this analysis is that it accounts at once for what might be called explanatory evasion—cases such as (from Achinstein [1975]) an aide responding to a journalist's 'Why did Kissinger shake hands warmly with Le Duc Tho?' by wilfully ignoring the obvious intended contrast-class and replying 'Because Kissinger, not Haig, is the chief negotiator'—without Achinstein's recourse to restructured paraphrase of the proper names.

Van Fraassen, and others who have advocated such an account of explanation, continue to presume that in each case the explanandum is a single proposition, the contrast-class being merely a context-determined aspect of the pragmatics of explanation. A much smoother account of explanation can be derived from importing the contrast-class into the explanandum. One version of this idea runs as follows.

In most, though not necessarily all, of the cases where we seek explanations we seek them not of one proposition *P* but of the ordered pair $<P, Q>$ (*P* rather than some of its contraries *Q, R, S* . . .). These propositions could be generalizations or statements about individual events, but are usually observational rather than theoretical statements. The available evidence points towards *P*'s being true in our actual world. We have formulated a theoretical statement T_p which gives reasonable grounds for expecting *P* (and hence not *Q*)—that is, T_p is statistically relevant to *P*. Now consider the modifications of T_p into theoretical statements $T_q, T_r, \ldots$ which give grounds for expecting *Q*, *R*, . . . rather than *P* while leaving everything else much the same. If some T_i is, in some sense, a preferable theory to T_p then, relative to T_p, $<P, Q>$ is unexplained and in need of explanation. Otherwise, T_p explains $<P, Q>$.

The account would be incomplete without an account of what it is for one theory to be preferable to another. It is tempting to rely upon relative simplicity—if there are no T_i's simpler than, or equally simple as, T_p then T_p explains $<P, Q>$. But in placing so much weight on the concept of simplicity we are bound to run into difficulties about the apparently subjective character of relative simplicity. Why not, then, try to run into these difficulties instead in a context where we have an answer to such problems—that of subjective probability? So, we can argue that T_p explains $<P, Q>$ iff there are no T_i's with a greater or equal subjective probability, for us. Of course, on an objectivist interpretation of probability this move would render scientific explanation disturbingly personal (although committed objectivists *may* be able to offer a similar account in terms of degree of confirmation). But for a thoroughgoing subjectivist this is not a problem: scientific explanation comes to depend on the only probabilities there are (our personal degrees of belief), governed by the constraint of coherence and made intersubjective by the process of conditionalization as theories are tested sequentially.

Let us return to the three problems from which this section began.

(i) Reference-class choice

If we want to explain why Jones, who had a cold and took vitamin C, recovered within a week, we need, according to Salmon, to select the broadest reference-class to which this event belongs which is yet homogeneous. If the class of people with colds is homogeneous with respect to recovery within a week, then we need not narrow our view to the class of people with colds who took vitamin C. In this way we screen off factors not causally relevant to the event. But how do we know when we have the broadest homogeneous reference-class? Is it not possible that we have ignored factors which are causally relevant, since that causal relevance is part of the physics of our world rather than of the logic of explanation?

What is happening here is that these empirical factors are being compressed into the explanans because, on standard accounts, there is no room for them in the explanandum; Salmon is then forced to try to explain homogeneity in terms of logical structure of the explanans. But if we instead allow the explananda to be complex—ordered pairs $<P, Q>$ of, in this case, individual-event descriptions—the requirement for homogeneity attaches to the explanandum: we must ensure that Q is the conjunction of all, and only, the causally relevant contraries of P. And that is a matter of empirical investigation into probabilities within the reference-class corresponding to each contrary.

(ii) Sufficiency for explanation

That T_p explains $<P, Q>$ requires that T_p be sufficient for the explanation of $<P, Q>$; it may very well not be sufficient for the explanation of $<P, R>$. This enables us to deal with such problems as the well-known paresis example. On a theory T_p which asserts that paresis develops in a small percentage of those with latent syphilis and in none of those who are unsyphilitic, the explanandum $<P, Q>$ where $P = $ 'A has paresis and is syphilitic' and $Q = $ 'A has paresis and is not syphilitic' has the same prior and posterior probability; whereas the explanandum $<P, R>$ where $P = $ 'A has paresis and is syphilitic' and $R = $ 'A does not have paresis and is syphilitic' has different prior and posterior probabilities. Hence T_p explains $<P, R>$ but not $<P, Q>$.

(iii) Asymmetry

What of cases such as the notorious flagpole example? The height of a pole and the length of its shadow are statistically relevant to one another (we should resist saying that one can be deduced from the other—theories such as that light travels in straight lines, to which we might, if we can, assign high probability but of which we are not certain, are involved) but we do not explain the pole's height by the shadow's length. We can concoct stories which use shadow length to explain why an atypical pole was built to a certain height but that goes no way toward accounting for the general asymmetry. That can be explained by the distinction between treating a proposition as one of the ordered pair $<P, Q>$ and as a statement T_p forming part of a theory. We more readily include statements of the pole's height than the shadow's length among our T_p statements because they represent more readily manipulable variables, more suited to rendering theories testable. No logical barrier exists against some T including statements of the shadow's length: the general asymmetry is pragmatic in origin.

These illustrations will, I hope, have confirmed what the proposed definition of explanation has suggested: that explanation is not an all-or-nothing matter. How powerful an explanation is will depend partly on how carefully we construct our Q-conjunction of causally relevant contraries; partly on the correct choice of explanandum so that T_p raises its, not another, probability; and partly on the testability of T_p independently of P (which is why 'P, because P' is such a poor explanation). It is not difficult to build a measure of explanatory power into the definition. In the possible world (or belief-set, or epistemic state) containing

P and T_p, since T_p gives grounds for expecting P rather than any of its contraries, $p(P)$ is greater than in any world I containing P and a T_i incompatible with T_p—remember that, though T_i gives grounds for expecting I rather than P, it is not incompatible with P: so there are such worlds. Then we could take $\{p(P)$ at the T_p-world–max$[p\ (P)$ at a T_i world$]\}$ as the degree to which T_p explains P. An explanatorily powerful explanation need not, however, be a good one—consider 'P, because P': a point to which I shall return.

(b) *What has Explanation to do with Conditionals?*

One view of this relation will not do, even though it seems to have attracted Goodman, Hempel, Reichenbach, and many others: that explanations rely upon laws and laws can be characterized by the derivability from them of counterfactual conditionals. First, explanations need not involve laws: theoretical statements T_p can take the form of singular propositions. Why did Adam eat the apple rather than a banana? A perfectly respectable explanation is that there were no bananas in Eden—but we could hardly call that a law. Secondly, we cannot simply claim (as e.g. Chisholm [1955] or Mackie [1973] do) that laws are counterfactualizable while mere accidental generalizations are not. It is a mere accident that all Smith's three dogs are white; but we can conclude from it that if that dog had been one of Smith's three dogs, it would have been white. What we cannot conclude, which we could if we were dealing with a law, is that if that dog had been Smith's fourth dog it would have been white. It is necessary to distinguish what have been called 'ampliative' from 'non-ampliative' counterfactuals. But we cannot use this as a reductive criterion of lawlikeness—for to say that a counterfactual holds ampliatively is to presuppose that it is supported by a projectible law.

These considerations suggest that the concept of law cannot form the link between explanation and conditionals. Yet there are several important respects in which the two are alike.

First, both most frequently function as answers to why-questions. 'Why have you invited him rather than me to your party?' receives the same answer in 'Because I expect(ed) him to bring his wife whom I want(ed) to meet' as in 'If he comes/came/had come, he will/would/would have brought his wife'. Notice that, when this resemblance is at issue, the tense and mood of the conditional are often unimportant.

Secondly, in both cases the demands to which we are responding

usually leave open a wide range of responses of which one is determined as appropriate by contextual factors.

Thirdly, the distinction between ampliative and non-ampliative conditionals closely parallels that between predictive and non-predictive
explanations, and in each case we prize the former type more. Everyone
in the room voted for the motion while the president was outside; we
know that if the president had been one of the people in the room she
would have voted for the motion, but that is uninteresting; what we
want to know is whether, if the president had been in the room as well
as the rest of us, she would have voted for the motion. Apart from
laboratory cases, no one catches malaria except those bitten by the
anopheles mosquito; currently, such mosquitoes only inhabit hot swampy
regions; so no one catches malaria except people in hot swampy regions. But location provides an inferior explanation. If we no longer
knew where mosquitoes were rife, we could still predict that no one
would catch malaria except those bitten by them, but we could say
nothing interesting about the malaria incidence in different regions.

I suggest that the resemblance is close enough so that we can treat
a large class of ordinary-language conditionals not, as has been suggested by Mackie [1973], as condensed arguments but as condensed
explanations. Of course there may well be if-sentences which do not fit
this pattern—Austin's example 'there are biscuits on the sideboard if
you want them' comes to mind—but most indicative, subjunctive, and
counterfactual conditional statements other than those embedded in
questions and commands seem to fit this model. The shared structure
of such conditionals would, then, be 'If T_p, then $<P, Q>$' where tense
and mood presuppose or conversationally imply the truth-values of T_p,
P and Q. Counterfactuals imply that T_p is false (in the actual world) and
leave the truth-value of P open, or less strongly imply it to be false;
either T_p explains P or not-T_p explains not-P. Indicative conditionals
have the same structure except that all the truth-values are left open.

Consider some examples.

1. 'If he comes to the party he will bring his wife.'

   His presence at the party (truth-value open) explains his wife's
   presence (truth-value open).

2. 'If he had come to the party he would have brought his wife.'

   His absence (implied) from the party explains his wife's (weakly
   implied) absence.

3. 'If he had been at the meeting he would have enjoyed himself.'

   His absence (implied) explains his (very weakly implied) not enjoying himself.

4. 'Since he came to the meeting he enjoyed himself.'

   His (stated) presence explains his (stated) enjoying himself.

The comparison between (3) and (4) is particularly interesting. Sentences constructed with 'since' are not, strictly, conditionals; but, as
Goodman [1955] pointed out, they resemble counterfactuals so
strongly—apart from attributing truth rather than implying falsehood—
that one might call them 'factual conditionals'. The parallel brings out
the viability of treating conditionals as explanations just as, less un-
expectedly, since-sentences are.

This strong suggestion about the nature of conditionals runs into
three immediate objections.

(i) Surely in explanations we take it that both explanans and explanandum are true? That claim can be disputed: Bohr's atomic theory was
thought to explain the hydrogen spectrum, but the theory has since
proven inadequate to cover other phenomena and the spectrum was
inaccurately measured—did the theory then cease to be an explanation
(rather than coming to be seen as a poor explanation)? But even if we
grant the claim, all that implies is that conditionals represent condensed
explanations of possible- rather than actual-world, phenomena.

(ii) Surely we cannot do justice to the logical force of 'if . . . then'
by explicating it in terms of a theory of explanation based on nothing
stronger than probabilistic relevance? Actually, it is just this distance
from logic which is one main advantage of this analysis. With the
exception of Quine [1950: 21]—'the problem of contrafactual conditionals belongs not to pure logic but to the theory of meaning or possibly the philosophy of science'—philosophers have tended to be lured
by their awareness that, in a formal logic extracted from everyday
reasoning, ⇒-connectives can be extracted from if-sentences, into treating if-sentences as loose applications of some formal calculus. This is
as true of those who, like Lewis [1973], Stalnaker [1968], or Adams
[1975], attempt to establish alternative logics of conditionals as of those
who adopt a materialist or logical powers stance. On the view I am
advancing, only a minority of if-sentences are used with any commitment to the antecedent's entailing the consequent.

(iii) The strength of this view can be seen clearly from what might at first seem like another objection to it. The statement 'If Imran makes a century, Pakistan will win the Test' is a condensed explanation of why, in the possible world in which they win, Pakistan will win—because Imran will make a century, which is statistically relevant to a Pakistan victory. But might someone not also believe that, even if Imran does not make a century, Pakistan will win? Surely he should not be indicted for advancing or believing two conflicting explanations of a Pakistan victory? Not in general, perhaps, but in some contexts he might. You interrupt my cricket-watching to opine that if Imran succeeds, Pakistan will win. I agree. You go on to say that if Imran fails, Pakistan will win. I become slightly annoyed. And quite reasonably, because a device exists in ordinary language—'whether or not Imran . . .'—to indicate that one is denying relevance of the antecedent to the consequent. The if-sentence asserts statistical relevance; that is why, when we believe there is none, we use 'whether' in preference to uttering the contrasting pair above, and why we expect, if we utter only one of them, to be taken to assert a connection between antecedent and consequent. Analysis in terms of condensed explanation helps account for this element of usage.

(c) *Conditionals and Probability*

The last section has already, I hope, illuminated some of the structure of conditionals. The treatment of conditionals as explanations should also shed light on some long-standing problems which can usefully be seen as parallel to the difficulties for theories of explanation explored in (*a*).

(i) *Relevance (cf. reference-class choice)*

Treated as material conditionals, both 'If England is part of Europe then snow is white' and 'If England is part of Europe then French and German foreign policies affect England' are true. Why is the latter interesting to us and the former not? Why do we rely upon some counterfactuals to guide our behaviour rather than others? If our view of conditionals is dominated by notions of them as arguments, or part of arguments, then we may be driven in answering these questions into challenging classical conceptions of validity and insisting on the need for relevance logic. But, if we treat conditionals as explanations, we can answer such questions with less drastic commitments. Recall Salmon's example of Jones, the sufferer from a cold. That a week has passed (given that colds generally last a few days only) explains the fact that Jones's cold has gone. That Jones took vitamin C is thus irrelevant to explaining why his cold has gone after a week rather than remaining—though it could be relevant to explaining why his cold has gone in a day rather than a week. That Jones is married is explanatorily irrelevant to both explananda. So, even if we would assign the truth-value True to 'If Jones is married, his cold will disappear in a week', treating it as material conditional, it is of no value as an explanation; 'if a week has passed, Jones's cold has gone rather than remaining' is an adequate explanation; 'if Jones took vitamin C, his cold will go in a day rather than a week' may be an adequate or inadequate explanation, but if we have grounds for expecting taking vitamin C to be relevant to the consequent it will be a conditional of interesting content—it will have some explanatory power.

(ii) Competing Counterfactuals (cf. sufficiency for explanation)

There is a well-known problem with mutually rebutting counterfactuals which both seem equally licensed, e.g.

'If Bizet and Verdi were compatriots, Bizet would be Italian rather than French' but
'If Bizet and Verdi were compatriots, Verdi would be French rather than Italian'.

Take the first of these. The prior probability that Bizet is Italian is (very close to) 0. That Bizet and Verdi were compatriots confers on it an increased posterior probability iff Bizet and Verdi were Italian compatriots, when the posterior probability is 1. Otherwise, their being compatriots is statistically irrelevant to Bizet's being Italian: the conditional is insufficient as an explanation. Exactly similar considerations apply to the second counterfactual. As Rescher [1961] points out, contextual ambiguity of the antecedent leaves us unable to choose between the two statements; what he fails to see, which this analysis makes clear, is that no choice is forced upon us—neither counterfactual is an assertion we would be justified in making, unless context supplies the needed statistical relevance.

(iii) Contraposibility (cf. Asymmetry)

Asserting some counterfactuals appears to commit us to accepting the contrapositive: from 'If he had been in the room he would have voted for the motion' we infer 'If he did not vote for the motion, he was not in the room'. But, as Goodman [1955] first pointed out, there are cases where we assert a conditional yet reject its contrapositive: an example (from Stalnaker [1968]) is the case of someone who, believing that North Vietnam will negotiate only if the USA agrees to complete withdrawal, asserts that 'If the USA stops the bombing, North Vietnam will not negotiate' while denying that 'If North Vietnam is negotiating, the USA will not have stopped the bombing'. We can resolve this difficulty in a parallel manner to that of asymmetry of explanation. Even if two statements are mutually statistical-relevant, it does not follow that each is equally good as an explanation of the other: taking a statement as the theory which explains $<P, Q>$ involves us in assessing its place in the web of our theoretical beliefs, whereas taking it as the P does not. Further, if T_p is statistically relevant to P, not-P need not be statistically relevant to not-T_p.

These three parallels with explanation suggest one way to deal with the problem of the meaning of probabilities of conditionals. I have argued for treating $A \rightarrow B$ as a condensed explanation of the form 'if A, then B rather than C'. As (*a*) indicated, each such explanation can be assigned a degree of explanatory power. The most natural way to interpret $p(A \rightarrow B)$ is, it seems to me, as equivalent to that degree of explanatory power. This will then provide us with an account of the acceptability of explanations which properly respects projective standards other than probabilistic coherence (superiority of one explanation to its competitors, 'surprise' value, etc.) while yet incorporating the constraints of projective probability. And it will provide us with an account of the assertibility conditions of conditionals which, since they share some of these explanatory standards and involve, while not reducing to, probabilities, allows us to see how those conditionals can (but need not) come, via projection of those standards, to possess truth-values.

4.4. PROJECTIVE JURIDICAL PROBABILITY

In parallel with the previous sections, our theory of probability can tolerate 'Baconianism' in the judicial context, though only if its pretensions there are localized: Cohen's claim that Baconian probability there displaces Pascalian must be resisted. It would be impossible to do justice here to the array of arguments L. J. Cohen has assembled to support that claim; all that can be attempted is to outline how this analysis of orders of probability, resiliency, and weight might apply to some of the main points. (Of course, as I suggested in Chapter 1, some counter-arguments to Cohen need not rely on these notions—they might be based on consequentialist considerations, for example.)

Cohen's general claim is that, while mathematical probabilities enter into judicial processes in various ways (e.g. in the presentation of statistical evidence), they do not bear upon juridical proof where it relates directly or mediately to ultimate issues, neither do they, nor should they, constitute the criminal or civil standard of proof—all of which are a matter for Baconian judgement. The notion of weight is central to this claim: he argues that trials, in an adversarial system, should be seen as contests of case-weight rather than attempts to persuade the trier of fact to allocate, on a basis of Pascalian probability, some measure of case-merit. It follows that neither of the two main standards of proof in Anglo-American law—proof beyond reasonable doubt and proof on the balance of probabilities—can be interpreted as given by mathematical probability (henceforth, p_m) without serious anomalies emerging. In his [1977] he put forward (apart from problems about inference upon inference which it seems he no longer wishes to rely upon) five such problems, which I shall consider in turn:

(a) that none of the generally accepted criteria of p_m could be applicable to evaluating juridical proof;

(b) that Pascalian explanations of the effect of testimonial corroboration involve, inappropriately, an assignment of positive prior probabilities;

(c) that proof beyond reasonable doubt cannot just be proof to a high level of p_m;

(d) that there is a difficulty with the conjunction rule for p_m's in civil cases;

(e) that there is a difficulty with the negation rule for p_m's in civil cases.

(a) For the strong subjectivist, the only criterion of p_m is, ultimately, rational betting. Cohen complained [1977] that that is inappropriate for two reasons: forensic matters deal with past events where all relevant evidence is assumed to be in, so bets would be unsettlable, and, since stake size generally affects betting-behaviour, we cannot make much of stakeless betting. He now (post-[1986]) seems to accept that these difficulties may be overcome, though at the cost of demanding extremely sophisticated introspection from jurors as to how they would bet in imagined circumstances. The real problem, he now argues, is that 'on a subjectivist interpretation each person's judgment of the probability of a particular proposition on given evidence does no more than describe that person's state of mind' [1989: 64]—when, of course, it should describe the case-strength. But only a very naïve subjectivism will fall foul of this criticism; as we have seen, SCS does establish projective standards of rationality for partial beliefs such that to claim that some proposition is probable is not to describe a belief but to express it, with the implicit claim that it meets such standards and with a commitment to its corrigibility if it does not. (Of course, it is important in the judicial context, as elsewhere, that treating projective probabilities thus should be compatible with being able meaningfully to embed probability-assignments within indirect contexts, within conditionals, and so on. This is a general challenge for quasi-realism which, as I shall indicate later, I think can be met quite generally; but additionally, the discussions of §2.4 and §4.3 should go some way towards showing responses to it which are specific to probability-judgements.) Furthermore, with the extension of coherence proposed here SCS now includes under such standards p_2 as well as p_1, and judgements of weight of evidence and resiliency—the leading idea is that we now have available p_2 criteria as well as p_1 ones. In judicial contexts, where it is rare to have exchangeability, these criteria will not normally lead directly to precise numerical values (as I pointed out in §2.4); but it would surely be a mark of something wrong with the theory if they did.

(b) Cohen is right in claiming that testimonial corroboration will not raise the probability of a particular verdict unless non-zero prior probability of that verdict is admitted, and he would be right to protest that such a non-zero prior is legally inadmissible—if p_1-values exhausted probability-judgements. On my view, jurors should and do start a case with any (set of) coherent priors (so long, in direct opposition to Cohen's view, as none is zero) but with a uniform p_2-distribution—that is, no especial commitment to any of these odds: there is no particular preferred prior p_1-value of guilt, rather than that prior p_1 having to be zero. This is judicially quite admissible: the p_2-dimension allows another way of representing having an open mind. Such situations are comparable to that of Miss Julie, discussed earlier, previewing the outcome of Match B where at first she knows nothing at all about either contestant. Whatever coherent set of prior p_1-values one has assigned to go with the uniform p_2-distribution, provided they include relevant dyadic as well as monadic judgements, as soon as evidence appears Bayesian conditionalization will come into play. And then there is no difficulty in seeing how the evidence affects p_1 while its weight affects p_2, nor of seeing how corroboration of testimony and convergence of circumstantial evidence can raise p_1-values.

(c) One cannot, Cohen claims, interpret 'beyond reasonable doubt' simply as meaning some high level of p_m. How could we say precisely—judges never try to—how high that level was to be? Instead what is needed, he argues, is some measure of how the evidence bears on the alleged guilt of the accused considered as individual, not as a member of a statistical reference class—so that p_m is an inappropriate measure. But it seems to me far from obvious that that is what a court requires: consider the use of genetic 'fingerprinting', for example, which is increasingly eroding former exclusions of expert evidence on ultimate issues. Still, the fact that English and American courts are beginning to express some unease about such methods lends at least some support to Cohen's claim that 'beyond reasonable doubt' does not just mean 'possessing a high p_1-value'. However, it does not follow that the absence of numerical guidance in judges' directions is to be taken as urging the jury that no particular high p_m is sufficient for conviction rather than as just avoiding specifying a precise line of demarcation. A refinement does need to be made to incorporate p_2 but it can be done just as in the civil case—see (e) below.

(d) If the civil standard were simply $p_m > 0.5$, then in situations where one party's case rested on a number of independent elements each of these would have to be established at a much higher level for their conjunction to remain above that standard. This, Cohen argues, is contrary to actual normal judicial practice. This argument is, I believe, fully met by developing Blackburn's [1980] point that such conjunction effects are not part of normal practice simply because truly independent elements of a case are so rare; but that when they exist, then the conjunction effect does yield a perfectly reasonable criterion of decision—accept the conjunction if its probability is better than $\frac{1}{2}$ and one which, so far as I can see, there is no clear evidence that triers of fact in either adversarial or inquisitorial systems fail to respect.

(e) The dificulty about negation comes out most vividly in Cohen's example [1977: 75] of the rodeo gatecrasher. Here (although Cohen is rather too ready to ignore the plight of the organizer, who, after all, is entitled to his day in court too) it can scarcely be adequate to take proof on the balance of probabilities as just $p_m > 0.5$, even if it is substantially greater. A clue to the correct analysis here is seen in the fact that English courts do sometimes take account of what Denning, in *Bater* v.

Bater (1951), called 'degrees of proof within that [civil] standard': a recent immigration case, *Khawaja* v. *Secretary of State* (1984), had the Law Lords speaking of 'the flexibility of the civil standard of proof' in satisfying a court that 'the facts which are required for the justification of the restraint put on liberty do exist' (this was, unusually, a case where the civil standard was being applied while loss of liberty was at stake). (These cases are cited in Cross and Tapper [1985: 142–3].) 'Degrees within a standard' could mean nothing if only thresholds of first-level p_m were involved—but nor could it if only thresholds of Baconian probabilities were involved. Evidently, what is needed here, and what such comments are groping towards, is the notion of a two-dimensional measure: we want $p_1 > 0.5$ and resiliency of p_1 to be high compared to $p_1 - 0.5$ (so that the p_2-distribution is concentrated above 0.5). In general it is enough in such cases that the α of greatest resiliency be such that res of $p_1(r)$'s being α is large compared to $\alpha - 0.5$, provided that w is acceptably large: variation in how large a resiliency and w-value are demanded is what is meant by degrees within the same standard—and that is equivalent to variations in requisite degrees of p_2 while p_1 is above 0.5. (In the criminal case just the same applies with $p_1 > n$ for arbitrarily large n.) This is the judicial practice here. If a justification of that practice is sought, we can find it by noting that the purposes for which trials are held are not satisfied merely by ensuring that more cases are decided correctly than incorrectly (as relying only on p_1 would ensure); stability of decisions, avoidance of revocation of penalties—which are what the high p_2 guards against—are necessary to our confidence in the cathartic as well as fact-finding effectiveness of judicial process, and justifiable on such consequentialist grounds. Both the reliability and the dramatic power of the process demand that evidence should be judged sufficient to convict only if it meets appropriate standards of first-order probability and of weight, taken as a function of second-order probability.

Conclusion

LET me now try to make explicit some connections and uncover some of the underpinnings of the projectivist theory which have perhaps remained somewhat obscured until now. A good way to start on this is, I think, to return to the seven questions asked of personalist theories in the Introduction: questions which arose out of comparisons with the objectivist alternatives. I consider these in (*a*) below. My responses to these questions will, of course, share the quasi-realist tenor of the whole of this work.

But I also think it worth while to go on, in (*b*), to ask how much quasi-realism has contributed to our understanding of probability. To avoid too much repetition here of what has been said in the last two chapters, I propose, in some necessarily brief remarks, to concentrate on how it enables us to avoid crude physical versus relativist dichotomies and on how it provides a solution to the criticisms raised by Vickers [1988]. Finally, I turn my gaze outwards and ask what the theory of projective probability (PP) does for the overall quasi-realist programme.

(a) *Questions for a Projectivist Theory*

(i) *'Are probabilities simply in the mind rather than in the world?'*

PP obviously relies directly on a quasi-realist approach in its response to such a question. Probabilities, on the PP account, are the projections onto the world of partial beliefs subject to certain idealizations and constraints. But they are not mind-dependent in the objectionable sense that they are what they are because you, or I, or anyone thinks so. Compare here moral projectivism. 'Is the wrongness of torture simply a sentiment in the mind?' On a quasi-realist view, there need be no error involved in claiming that it is a fact, or that it is objectively correct, or that we might come to appreciate or discover that torture is wrong. Realistic talk of values is quite in order for the quasi-realist: it is the end result of forming evaluations by applying certain standards and constraints to the projection of our sentiments. Similarly, as Chapter 3 has argued, PP accepts the legitimacy of surface realism in our talk of probabilities while carrying out the task of demonstrating their subjective source. And, similarly, it comes to query the clear-cut and final nature of the objective–subjective distinction being presumed.

(ii) *'Are probabilities just actual beliefs, or does PP offer a normative theory of partial belief?'*

§1.5 provides, I think, a full answer. It sees this distinction as taking it for granted that the only shape a normative theory of human action can take is formed on what Cohen [1981] has called a 'preconceived norm model' rather than a 'norm extraction model'; and, on the latter model, PP can both be an idealized description and have regulative force.

(iii) *'Which is the more basic concept—partial belief or probability?'*

As I suggested earlier, Day's [1961] point here is that a theory which sets out to interpret probability in terms of partial belief is in danger of explaining the obscure in terms of the more obscure. But, as Chapter 2 made clear, the concept of partial belief—if it is supposed to extend beyond dispositions to act in ways measurable by willingness to bet— plays no essential part in the developed theory. None the less, the thought behind this point can resurface in a more subtle objection. Is PP offering a reduction of probability to dispositions to act? And, if so, is it reducing what should properly be a property of the contents of judgement to a property of acts of judgement? I take up this issue in (*b*) below.

(iv) *'Is no coherent belief-set ever more reasonable or justified than another?'*

Answering this question has been my central concern in Chapter 2, which has argued that the effects of coherence upon belief-sets are much more powerful than might at first be thought. Suppose that you, living in Manchester, assign in advance today a probability-value of 0.001 to each of the events of rain tomorrow, rain the following day, and so on for the next 365 days. You also believe that each day is much the same as another as far as the risk of rain goes, so you treat these events as exchangeable. Then, as we saw, if you conditionalize on the evidence as the days pass, coherence alone forces you to alter that value towards the observed precipitation frequency. More than that: if initially you had no particular confidence in any probability-value, or you were very confident of a value which differed from the frequency, if you conditionalize on the evidence then coherence alone will concentrate that second-order confidence onto a probability-value more in line with actual frequency. So coherence is not the relaxed constraint which it might seem to be. But when we come to ask for justification of one of your possible belief-sets against another, the key clause can readily be seen to be 'if you conditionalize on the evidence'. And here PP rightly does not attempt a justification. A Pyrrhonian inductive sceptic will, presumably, not conditionalize at all; certainly, some answer is needed to such scepticism—we need a projectivist account of our inductive habits too. But any attempts to build such an account into a theory of probability which also is an interpretation of the formal calculus run into horrendous problems. PP avoids that foolhardiness.

(v) *'How and why is so much actual consensus about probabilities possible?'*

PP has a direct answer here, obtained by combining §1.5 and Chapter 2: the convergence mechanism outlined above is an idealization of features in our actual practice which effect and regulate such consensus.

(vi) *'Does the central place of betting-measures in PP mean that the theory presupposes a concept of utility?'*

Unfortunately, yes—as I conceded in §2.4: though only to the extent that before we can bet there must already be in place some criterion of utility for some fragment of our preferences. I am not fully happy with this answer, however, since it threatens to reimport into PP at least some of the problems (§1.1) afflicting preference-based theories. Perhaps it might instead be possible to evolve a conception of coherence as a rationality constraint which precluded conflicting commitments emerging from judging differently something which is one single option variously described. Ramsey [1931: 182] may have been hinting at this in offering what looks like a non-utility-based definition of probabilistic inconsistency (i.e. incoherence) just before the Dutch book version:

If anyone's mental condition violated these laws [namely the axioms of probability], his choice could depend on the precise form in which the options were offered him, which would be absurd. He could have a book made against him by a cunning bettor and would then stand to lose in any event.

But I am as yet unsure how this idea might be developed.

(vii) *'How can it be in order to form probability-judgements in circumstances where no bet is settlable?'*

I have given piecemeal answers to this question in §1.2 and Chapter 4 which provide, I think, a patchy but not wholly inadequate response.

I suspect that a fully adequate response may be possible only with the development of an answer, along the lines sketched earlier, to (vi).

(b) *The Mutual Debts of PP and Quasi-Realism*

I have claimed earlier that PP can sustain its pretensions to being a successful univocal interpretation of probability even if one does not take a quasi-realist stance on its achievements. But it does undoubtedly sit very comfortably with a quasi-realist perspective.

(i)

Quasi-realism helps us to locate the space for our theory between strongly realist and crudely relativist theories. The latter viewpoint is exemplified in the personalism of writers like Borel, and its bankruptcy is exposed by their inability to do justice to the objectivism of probability usage. Borel finds himself trying to give some sense to the notion of objective or intersubjective probability as 'the probability which is common to the judgements of all the best informed persons, that is to say, the persons possessing all the information that it is humanly possible to possess at the time' [1964: 28], but with no supporting theory as to what might bring this consensus about. The former viewpoint finds expression in a recent (unpublished) paper by Deutsch [1989], where he argues that, unless probabilities are factual attributes of physical events they could play no part in fundamental physical theories which aim at 'describing reality'. Quasi-realism enables us to see that probability statements can be factual, can 'describe reality', without probability being a physical attribute—any more than morality or modality must be physical attributes for moral or modal statements to be factual.

(ii)

Are there not better arguments than Deutsch's for taking probabilities to be part of the physical world? Mellor [1982] argues that actions based upon probability-judgements could only be justified if probabilities can enter into causal interactions, if there are, beyond degrees of belief, chances in his propensity-theory sense; and 'a probability which has causes is a part of the physical world' [1982: 105]. But quasi-realism is quite able to treat causation after a Humean manner, so that causal connections are rightly said to be 'objective' or 'physical' while yet being projections of relations among our judgements. So it has no difficulty in pursuing a similar line with causal probability.

(iii)

Quasi-realism supplies PP with an alternative to meekly accepting being labelled a 'non-cognitivist' theory. That label generally presupposes a distinction between cognitive or truth-conditional commitments such as knowledge claims and non-cognitivist or non-truth-functional commitments such as expression of one's distaste or binding oneself by promises. Following Wittgenstein, we should be wary of supposing that there are cognitive commitments which somehow float free of consensual norms and practices. But in any case the distinction mislocates the issue of what is needed to ensure that probability-judgements are not implausibly relativist. That issue is resolved once we note that the coherence constraint gives rise to a calculus which ensures that what we can correctly say, what judgements we can rationally project, is not entirely in our own hands: it imposes standards upon our probabilistic commitments which ensure that these stances end up with truth-conditional expression.

(iv)

Vickers [1988: 231] inveighs against subjectivism's

refusing to question certain anachronistic, strong, and inadequate presumptions about the nature and form of judgement. Indeed, as far as the question of the nature of probable knowledge is concerned, much of the epistemological discussion of this century . . . might never have taken place. If there has been recent progress in philosophy, the development of a general and recursive theory of propositional structure and the concomitant translation of features of the act of judgement into recurring and embedded parts of the object of judgement is certainly an instance of it. . . . As concerns probability, however, only empirical frequentism and propensity interpretations have made any effort to shift probability from the act to the object of probable judgement, and neither of these has provided or could easily provide a thorough account of embedded probabilities. Carnapian logicism and subjectivism are hostile or indifferent to the question of embedded probabilities.

Vickers is presenting PP with a dilemma here. So long as its account of probability is based upon acts of judgement, it will never escape an anachronistic psychologism. If, however, it attempts to shift its base to the objects of judgement, the theory will be obliged to deal with recursion and embedding of probabilities; and it will not be able to cope with these, Vickers claims, because such features—which are the mark of the logical—can only be handled by a refined logicist theory, one version of which he goes on to articulate.

There is both a drastic and a less drastic response to this argument. The drastic response would be to attack Vickers's elevation of content over formation and expresssion of judgements and attitudes. But the problem with this line is that it renders the defence of PP dependent upon a global position on the nature of thought, language, belief, and knowledge. Fortunately, quasi-realism provides a less drastic response which adequately meets Vickers's strictures: accepting that PP should be a theory of the objects of judgements which may involve recursion, but denying that only a logical-relation theory can smoothly handle recursion. §2.4 has already shown how to handle one type of recursion, the embedding of probabilities within higher-order probability-judgements. Other forms of embedding—within the scopes of truth-functions or of quantifiers—raise precisely the same problem for PP as the notorious Frege-Geach argument (see Geach [1964]) about indirect contexts raises for expressive theories in ethics. Blackburn ([1984], and subsequently—see [1993]) has shown how a projectivist moral theory can meet this argument by explaining the possibility of, and providing a semantics for, embedded commitments based upon internal consistency standards. In a parallel manner—and rather more easily—PP's clear articulation of coherence, and its links to consistency, generate an account of recursion which in no way threatens the view of probabilities as projected subjective commitments.

(v)

Finally, let us turn briefly to survey what PP can do to repay some of its debt to quasi-realism.

First, the most obvious—though not on account of such obviousness any-the-less powerful—criticism of the general quasi-realist position is that it makes improper use of the notion of evaluations coming to earn realist-sounding expression provided they are projected according to 'proper' standards. The question arises—as it does with attempted analytic justifications of induction—whether 'proper' here means 'objectively correct', in which case realist assumptions are being smuggled in, or means 'internally coherent', when we have an anti- rather than quasi-realism with no closure of the realism–anti-realism gap. PP has shown that the only way to meet this criticism is to establish standards of propriety which are independent of factual vindication but which are not in the individual's own hands—not a matter of taste, in Hume's sense.

Secondly, the formidable problems which have arisen within PP in striving to establish that coherence is a sufficient constraint to account for all of what Mellor [1971] called 'the objectivism of usage' issues a warning that quasi-realism has as yet largely only identified and argued the possibility of a range of formidable tasks, still to be carried out.

Thirdly, the issues discussed in Chapter 4 especially raise important questions as to the drawbacks as well as the advantages of global quasi-realism, of bundling together issues of moral, modal, probabilistic, conditional, etc. commitment. Advantages include a stronger view of the nature of fact and of truth-functionality, of the relationship between a stance and an epistemic state; drawbacks include the temptation towards excessive optimism about the range of consequences of a quasi-realist approach.

Lastly, that PP is a genuine resolution of otherwise implacable theoretical problems should help quasi-realism to answer the charge that, if it succeeds, it can only do so at the cost of demonstrating that there never was a gap to be crossed between anti-realist starting intuitions and the realist endpoint. It would be altogether too hard on a theory which, surrounded by competitors arguing whether probability is a knot or a straight piece of string, succeeded in untying the complex knot, to accuse it of self-defeating success.

REFERENCES

ACHINSTEIN, P. [1975], 'The Object of Explanation', in S. Körner (ed.), *Explanation* (Oxford).

ADAMS, E. [1975], *The Logic of Conditionals* (Boston, Mass.).

ADLER, J. [1991], 'An Optimist's Pessimism: Conversation and Conjunction', in Eells and Maruszewski [1991].

AUMANN, R. J. [1976], 'Agreeing to Disagree', *Annals of Statistics*, 6: 1236–9.

AYER, A. J. [1957], 'The Conception of Probability as a Logical Relation', in S. Körner (ed.), *Observation and Interpretation* (London).

—— [1972], *Probability and Evidence* (London).

AYERS, M. R. [1968], *The Refutation of Determinism* (London).

BAYES, T. [1763], 'An Essay towards Solving a Problem in the Doctrine of Chances', *Philosophical Transactions of the Royal Society*, 53: 370–418.

BENENSON, F. C. [1984], *Probability, Objectivity and Evidence* (London).

BIGELOW, J. [1976], 'Possible Worlds Foundations for Probability', *Journal of Philosophical Logic*, 5: 299–320.

BLACK, M. [1967], 'Probability', *Encyclopedia of Philosophy* (New York).

BLACKBURN, S. [1973], *Reason and Prediction* (Cambridge).

—— [1980a], 'Opinions and Chances' in D. H. Mellor (ed.) *Prospects for Pragmatism* (Cambridge).

—— [1980b], Review of *The Probable and the Provable*, *Synthese*, 44: 149–59.

—— [1981], 'Rational Animal?', *Behavioural and Brain Sciences*, 4/3: 331–2.

—— [1984], *Spreading the Word* (Oxford).

—— [1993], *Essays in Quasi-Realism* (New York).

BOREL, E. [1964], 'Apropos of a Treatise on Probability', trans. in Kyburg and Smokler [1964], First pub. [1924].

CARNAP, R. [1950], *The Logical Foundations of Probability* (Chicago).

—— and JEFFREY, R. C. [1971], *Studies in Inductive Logic and Probability* (Berkeley, Calif.).

CARTWRIGHT, N. [1979], 'Causal Laws and Effective Strategies', *Noûs*, 13: 419–37.

CHISHOLM, R. [1955], 'Law Statements and Counterfactual Inference', *Analysis*, 15.

COHEN, J. [1972], *Psychological Probability* (London).

COHEN, L. J. [1970], *The Implications of Induction* (London).

—— [1977], *The Probable and the Provable* (Oxford).

—— [1979], 'On the Psychology of Prediction: Whose is the Fallacy?', *Cognition*, 7: 385–407.

—— [1981], 'Can Human Irrationality be Experimentally Demonstrated?', *Behavioural and Brain Sciences*, 4/3: 317–30.

—— [1982], 'Are People Programmed to Commit Fallacies?', *Journal for the Theory of Social Behaviour*, 12: 251–74.

—— [1986*a*], *The Dialogue of Reason* (Oxford).

—— [1986*b*], 'The Role of Evidential Weight in Criminal Proof', *Boston University Law Review*, 66: 635–49.

—— [1986*c*], 'Twelve Questions about Keynes's Concept of Weight', *British Journal for the Philosophy of Science*, 37: 263–78.

—— [1989], *An Introduction to the Philosophy of Induction and Probability* (Oxford).

—— [1991], 'Some Comments by L.J.C.', in Eells and Maruszewski [1991].

—— [1992], *An Essay on Belief and Acceptance* (Oxford).

CROSS, R., and TAPPER, C. [1985], *Cross on Evidence*[6] (London).

DARWIN, C. [1859], *On the Origin of Species by Means of Natural Selection* (London).

DAVIDSON, D. [1985], 'A New Basis for Decision Theory', *Theory and Decision*, 18: 87–98.

—— SUPPES, P., and SIEGEL, S. [1957], *Decision Making: An Experimental Approach* (Stanford).

DAWID, A. P. [1977], 'Invariant Distributions and Analysis of Variant Models', *Biometrika*, 64: 291–7.

—— [1982], 'The Well-calibrated Bayesian', *Journal of the American Statistical Association*, 77: 605–13.

—— [1985], 'Probability, Symmetry and Frequency', *British Journal for the Philosophy of Science*, 36: 107–28.

—— and DICKEY, P. [1977], 'Likelihood and Bayesian Inference from Selectively Reported Data', *Journal of the American Statistical Association*, 72: 845–50.

DAY, J. P. [1961], *Inductive Probability* (London).

DEUTSCH, D. [1989], 'Probability in Physics', *Oxford University Mathematical Institute* (mimeo) (Oxford).

DORLING, J., and MILLER, D. [1981], 'Bayesian Personalism, Falsificationism and the Problem of Induction', *Proceedings of the Aristotelian Society* supp. 55: 109–41.

DUMMETT, M. [1982], 'Realism', *Synthese*, 46: 55–112.

EELLS, E. [1982], *Rational Decision and Causality* (Cambridge).

EELLS, E., and MARUSZEWSKI, T. [1991] (eds.), *Probability and Rationality: Studies on L.J. Cohen's Philosophy of Science* (Amsterdam).

ELLIS, B. [1966], *Basic Concepts of Measurement* (Cambridge).

—— [1973], 'The Logic of Subjective Probability', *British Journal for the Philosophy of Science*, 24: 125–52.

—— [1979], *Rational Belief Systems* (Oxford).

FIELD, H. [1977], 'Logic, Meaning and Conceptual Role', *Journal of Philosophy*, 74: 379–409.

FINE, T. [1970], *Theories of Probability* (New York).

FINETTI, B. De [1931], 'Probabilismo', *Logos*, 163–219.

—— [1937], 'La Prévision, ses lois logiques, ses sources subjectives', *Annales de l'Institut Henri Poncaré*, 7 (trans. in Kyburg and Smokler [1964]).

—— [1972], *Probability, Induction and Statistics* (New York).

—— [1974], *Theory of Probability*, 2 vols. (New York).

GÄRDENFORS, P. [1988], *Knowledge in Flux* (Cambridge, Mass.).

—— and SAHLIN, N.-E. [1982], 'Unreliable Probabilities, Risk Taking and Decision Making', *Synthese*, 53: 361–86.

GEACH, P. [1964], 'Assertion', *Philosophical Review*, 74: 449–65.

GIGERENZER, G., SWIJTINK, Z., PORTER, T., DASTON, L., BEATTY, J., [1989], *The Empire of Chance* (Cambridge).

GOOD, I. J. [1962], 'Subjective Probability as the Measure of a Non-measurable Set', in E. Nagel, P. Suppes, and A. Tarski (eds.), *Logic, Methodology and Philosophy of Science* (Stanford).

GOODMAN, N. [1955], *Fact, Fiction and Forecast* (Harvard, Mass.).

HACKING, I. [1965], *The Logic of Statistical Inference* (Cambridge).

—— [1967], 'Slightly More Realistic Personal Probability', *Philosophy of Science*, 34: 311–25.

HANSSON, B. [1975], 'Explanations—of What?', *Stanford, Philosophy Dept.* (mimeo).

HORWICH, P. [1982], *Probability and Evidence* (Cambridge).

HOWSON, C., and URBACH, P. [1989], *Scientific Reasoning: The Bayesian Approach* (La Salle, Ill.).

HUME, D. [1739], *A Treatise of Human Nature* (London).

HURD, A. E. [1983] (ed.), *Nonstandard Analysis: Recent Developments* (New York).

JEFFREY, R. C. [1974], 'Preference among Preferences', *Journal of Philosophy*, 71: 377–91.

—— [1983], *The Logic of Decision* (Chicago) [1st pub. 1965].

—— [1984], 'De Finetti's Probabilism', *Synthese*, 60: 73–90.

KADANE, J. [1982], Comment on Dawid [1982: 610–11].

KAHNEMAN, D., and TVERSKY, A. [1972], 'Subjective Probability: A Judgment of Representativeness', *Cognitive Psychology*, 3: 430–54.

—— [1973], 'On the Psychology of Prediction', *Psychological Review*, 80: 237–51.

—— [1974], 'Judgment under Uncertainty: Heuristics and Biases', *Science*, 185: 1124–31.

—— [1979], 'On the Interpretation of Intuitive Probability: A Reply to Jonathan Cohen', *Cognition*, 7: 409–11.

—— [1982], *Judgment under Uncertainty: Heuristics and Biases* (New York).

KEMENY, J. [1955], 'Fair Bets and Inductive Probabilities', *Journal of Symbolic Logic*, 20: 263–73.

KEYNES, J. M. [1921], *A Treatise on Probability* (London).

KNEALE, W. [1949], *Probability and Induction* (Oxford).

KOLMOGOROV, A. N. [1933], 'Grundbegriffe der Wahrscheinlichkeitsrechnung', *Ergibnisse der Mathematik*, 2: 196–262.

KYBURG, H. [1968], 'Bets and Beliefs', *American Philosophical Quarterly*, 5: 54–63.

—— [1974], *The Logical Foundations of Statistical Inference* (Dordrecht).

—— [1975], 'The Uses of Probability and the Choice of a Reference Class', in W. Harper and C. Hooker (eds.), *Foundations of Probability Theory* (Dordrecht).

—— [1978], 'Subjective Probability: Criticisms, Reflections and Problems', *Journal of Philosophical Logic*, 7: 157–80.

—— [1981], 'Principle Investigation', *Journal of Philosophy*, 78: 772–8.

—— and SMOKLER, H. [1964] (eds.), *Studies in Subjective Probability* (New York).

LEHRER, K. [1972], 'Evidence and Conceptual Change', *Philosophia*, 2: 273–82.

LEVI, I. [1967], *Gambling with Truth* (New York).

—— [1970], 'Probability and evidence', in M. Swain (ed.), *Induction, Acceptance and Rational Belief* (Dordrecht).

—— [1978], 'Confirmational Conditionalization', *Journal of Philosophy*, 74: 730–7.

—— [1980], *The Enterprise of Knowledge* (Cambridge, Mass.).

LEWIS, D. [1973], *Counterfactuals* (Oxford).

—— [1976], 'Probabilities of Conditionals and Conditional Probability', *Philosophical Review*, 85: 297–315.

—— [1986], 'A Subjectivist's Guide to Objective Chance', *Philosophical Papers* (New York), ii. 83–132 [1st pub. 1980].

LINDLEY, D. [1982], 'The Bayesian Approach to Statistics', in J. T. de Olivera (ed.), *Some Recent Advances in Statistics* (New York).

LOGUE, J. [1987], 'Coherence, Convergence and Consensus', B.Phil. thesis (Oxford).

—— [1991], 'Weight of Evidence, Resiliency and Second-order Probabilities', in Eells and Maruszewski [1991].

LOPES, L., and ODEN, G. [1991], 'The Rationality of Intelligence', in Eells and Maruszewski [1991].

LUNTLEY, M. [1988], *Language, Logic and Experience: The Case for Antirealism* (La Salle, Ill.).

MACKIE, J. [1973], *Truth, Probability and Paradox* (Oxford).

MARTIN-LÖF, P. [1969], 'The Literature on von Mises' Kollektivs Revisited', *Theoria*, 35: 12–37.

MELLOR, D. H. [1971], *The Matter of Chance* (Cambridge).

—— [1980], 'Consciousness and Degrees of Belief', in D. H. Mellor (ed.), *Prospects for Pragmatism* (Cambridge).

—— [1982], 'Chances and Degrees of Belief', in McLaughlin (ed.), *What, Why, Where, When* (Boston, Mass.).

MILLER, R. W. [1987], *Fact and Method* (Princeton, NJ).

PEACOCKE, C. [1979], *Holistic Explanation* (Oxford).

PEIRCE, C. S. [1932], *Collected Papers of Chartes Sandes Peirce*, ed. C. Harts-horne and P. Weiss (Cambridge, Mass.).

PETERSON, C. R. and BEACH, L. R. [1967], 'Man as an Intuitive Statistician', *Psychological Bulletin*, 68: 29–46.

PHILLIPS, L., and EDWARDS, W. [1966], 'Conservatism in a Simple Probability Inference Task', *Journal of Experimental Psychology*, 72: 346–54.

POPPER, K. [1957], 'The Propensity Theory of the Calculus of Probability and the Quantum Theory', in S. Körner (ed.), *Observation and Explanation* (London).

—— [1959], *The Logic of Scientific Discovery* (London).

PRICE, H. [1986], 'Conditional Credence', *Mind*, 95: 18–36.

PUTNAM, H. [1963], '"Degree of Confirmation" and Inductive Logic', in P. Schilpp (ed.), *The Philosophy of Rudolf Carnap* (La Salle, Ill.).

—— [1975], 'The Meaning of "Meaning"', in K. Gunderson (ed.), *Language, Mind and Knowledge* (Minneapolis, Minn.).

QUINE, W. V. O. [1950], *Methods of Logic* (London).

RAMSEY, F. P. [1931], 'Truth and Probability', in *The Foundations of Mathematics and other Logical Essays* (London).

REICHENBACH, H. [1949], *The Theory of Probability* (Berkeley, Calif.).

RESCHER, N. [1961], 'Belief-contravening Suppositions and the Problem of Contrary-to-fact Conditionals', *Philosophical Review*, 70: 176–96.

ROBINSON, A. [1970], *Non-Standard Analysis* (Amsterdam).

ROSENKRANTZ, R. [1977], *Inference, Method and Decision* (Dordrecht).

—— [1978], 'Rational Information Acquisition', in L. J. Cohen and M. Hesse (eds.), *Applications of Inductive Logic* (Oxford).

RYDER, J. M. [1981], 'Consequences of a Simple Extension of the Dutch Book Argument', *British Journal for the Philosophy of Science*, 32: 164–7.

SALMON, W. [1967], *The Foundations of Scientific Inference* (Pittsburgh, Pa.).

—— [1978], 'Why Ask Why?', *Proceedings of the American Philosophical Association*, 51.

—— with contributions by R. Jeffrey and J. Greeno [1971], *Statistical Explanation and Statistical Relevance* (Pittsburgh, Pa.).

SAVAGE, L. J. [1954], *The Foundations of Statistics* (New York).

SEIDENFELD, T. [1985], 'Coherence, Calibration and Proper Scoring Rules', *Philosophy of Science*, 52: 274–94.

SHAFER, G. [1978], *A Mathematical Theory of Evidence* (Princeton, NJ).

—— [1983], 'A Subjective Interpretation of Conditional Probability', *Journal of Philosophical Logic*, 12: 453–66.

SKYRMS, B. [1975], *Choice and Chance* (Belmont, Calif.).

—— [1980a], *Causal Necessity* (New Haven, Conn.).

—— [1980b], 'Higher Order Degrees of Belief', in D. H. Mellor (ed.), *Prospects for Pragmatism* (Cambridge).

STALNAKER, R. [1968], 'A Theory of Conditionals', in N. Rescher (ed.), *Studies in Logical Theory* (Oxford).

TELLER, P. [1973], 'Conditionalisation and Observation', *Synthese*, 26: 218–58.

TOULMIN, S. [1950], 'Probability', in A. Flew (ed.), *Essays in Conceptual Analysis* (London).

VAN FRAASSEN, B. [1980*a*], 'Rational Belief and Probability Kinematics', *Philosophy of Science*, 47: 165–87.

—— [1980*b*], *The Scientific Image* (Oxford).

—— [1983], 'Calibration: A Frequency Justification for Personal Probability', in R. S. Cohen and L. Laudan (eds.), *Physics, Philosophy and Psychoanalysis* (Dordrecht).

—— [1984], 'Belief and the Will', *Journal of Philosophy*, 81: 235–56.

—— [1985], 'Empiricism in the Philosophy of Science', in P. M. Churchland and C. A. Hooker (eds.), *Images of Science* (Chicago).

VICKERS, J. M. [1988], *Chance and Structure* (Oxford).

VON MISES, R. [1957], *Probability, Statistics and Truth* (London) [1st pub. 1928].

VON NEUMANN, J., and MORGENSTERN, J. [1947], *The Theory of Games and Economic Behaviour* (Princeton, NJ).

WAISMANN, F. [1965], *The Principles of Linguistic Philosophy* (London).

WEATHERFORD, R. [1982], *Philosophical Foundations of Probability Theory* (London).

WITTGENSTEIN, L. [1975], *Philosophical Remarks* (Oxford).

INDEX